Brian R. Burg

Directing Carbon Nanotubes and Graphene

Brian R. Burg

Directing Carbon Nanotubes and Graphene

Fundamentals and Parallel Device Integration using Dielectrophoresis

Südwestdeutscher Verlag für Hochschulschriften

Impressum/Imprint (nur für Deutschland/only for Germany)
Bibliografische Information der Deutschen Nationalbibliothek: Die Deutsche Nationalbibliothek verzeichnet diese Publikation in der Deutschen Nationalbibliografie; detaillierte bibliografische Daten sind im Internet über http://dnb.d-nb.de abrufbar.
Alle in diesem Buch genannten Marken und Produktnamen unterliegen warenzeichen-, marken- oder patentrechtlichem Schutz bzw. sind Warenzeichen oder eingetragene Warenzeichen der jeweiligen Inhaber. Die Wiedergabe von Marken, Produktnamen, Gebrauchsnamen, Handelsnamen, Warenbezeichnungen u.s.w. in diesem Werk berechtigt auch ohne besondere Kennzeichnung nicht zu der Annahme, dass solche Namen im Sinne der Warenzeichen- und Markenschutzgesetzgebung als frei zu betrachten wären und daher von jedermann benutzt werden dürften.

Verlag: Südwestdeutscher Verlag für Hochschulschriften GmbH & Co. KG
Dudweiler Landstr. 99, 66123 Saarbrücken, Deutschland
Telefon +49 681 37 20 271-1, Telefax +49 681 37 20 271-0
Email: info@svh-verlag.de

Approved by: Zurich, Diss. ETH No. 19095, 2010.

Herstellung in Deutschland:
Schaltungsdienst Lange o.H.G., Berlin
Books on Demand GmbH, Norderstedt
Reha GmbH, Saarbrücken
Amazon Distribution GmbH, Leipzig
ISBN: 978-3-8381-2734-7

Imprint (only for USA, GB)
Bibliographic information published by the Deutsche Nationalbibliothek: The Deutsche Nationalbibliothek lists this publication in the Deutsche Nationalbibliografie; detailed bibliographic data are available in the Internet at http://dnb.d-nb.de.
Any brand names and product names mentioned in this book are subject to trademark, brand or patent protection and are trademarks or registered trademarks of their respective holders. The use of brand names, product names, common names, trade names, product descriptions etc. even without a particular marking in this works is in no way to be construed to mean that such names may be regarded as unrestricted in respect of trademark and brand protection legislation and could thus be used by anyone.

Publisher: Südwestdeutscher Verlag für Hochschulschriften GmbH & Co. KG
Dudweiler Landstr. 99, 66123 Saarbrücken, Germany
Phone +49 681 37 20 271-1, Fax +49 681 37 20 271-0
Email: info@svh-verlag.de

Printed in the U.S.A.
Printed in the U.K. by (see last page)
ISBN: 978-3-8381-2734-7

Copyright © 2011 by the author and Südwestdeutscher Verlag für Hochschulschriften GmbH & Co. KG and licensors
All rights reserved. Saarbrücken 2011

Document typeset by the author using LATEX 2_ε with the KOMA-SCRIPT document class scrbook. Text is set in Palatino. Figures were prepared with Matlab and Adobe Illustrator.

© Brian Burg, Zurich (2010).

What I want to talk about is the problem of
manipulating and controlling things on a small scale.

Richard P. Feynman (1918-1988)
in *There's Plenty of Room at the Bottom* on December 29, 1959.

Abstract

The dielectrophoretic integration of carbon nanostructures for parallel sensor assembly is reported in the present thesis. The aim is to provide a viable avenue to ensure continuous miniaturization toward nanoelectromechanical systems (NEMS), driven by cost reduction, enhanced device functionality, and improved energy efficiency. It is proposed to employ directed assembly of high-symmetry low-dimensional materials, while always allowing parallel integration. For this purpose, the electrokinetic framework of dielectrophoretic deposition devices is first developed, in order to understand and exploit the occurrence and interaction of the different underlying effects in capacitively coupled systems when moving nano-sized particles to desired locations. In the following, a method for dispersing surface-synthesized individual single-walled carbon nanotubes (SWNTs) in ultrapure and long-term stable aqueous solutions by low energy input is introduced, which is necessary for the successful high-yield dielectrophoretic deposition of these one-dimensional carbon nanostructures. Electrical characterization on 223 low-resistance devices evidences the high quality of the SWNT solutions, raw material and contact interface. Subsequently, the dielectrophoretic separation of individual metallic SWNTs from heterogeneous solutions is proven. Their simultaneous deposition between electrodes is confirmed by direct electric transport measurements. A threshold separation frequency of 188 MHz is extracted from a surface-conductivity model and a conductivity weighting factor introduced to elucidate the separation frequency dependence. To show the versatility of the introduced methods, the parallel integration of SWNT based piezoresistive pressure sensors is demonstrated. Carbon nanotubes are dielectrophoretically placed at the membrane edges, the positions of largest strain. Highest sensitivity of the long-term stable devices is achieved in the off-state of small band gap carbon nanotubes (SGS-SWNTs), reaching values as high as $S_0 \sim 0.25\ \Delta R/R/\text{bar}$, at a resolution better than 50 mbar, and power consumption of less than 40 nW. In the end, the dielectrophoretic integration of single- and few-layered graphenes, two-dimensional carbon nanostructures, is demonstrated. The above accomplishments allow new approaches to counter ongoing miniaturization efforts. Low cost solution based technologies are capable of assembling functional transducer elements up to 3 orders of magnitude smaller than the current state of the art while enhancing sensor resolution and significantly reducing drive current requirements.

Zusammenfassung

In der vorliegenden Doktorarbeit wird die dielektrophoretische Integration von Kohlenstoffnanostrukturen für deren parallele Sensorfabrikation untersucht. Ziel ist es dabei die Miniaturisierung von nanoelektromechanischen System weiterzuentwickeln, angetrieben von Kosteneinsparungen, erhöhter Funktionalität und reduziertem Energieverbrauch. Die gezielte Anordnung von Materialien hoher Symmetrie und niedriger Dimensionalität wird dafür in dieser Arbeit erforscht, unter Beibehaltung der parallelen Integrationsfähigkeit. Zu diesem Zweck wird zuerst das elektrokinetische Verhalten von dielektrophoretischen Ablagerungssystemen beschrieben, um die auftreten Effekte und deren Interaktion in kapazitiv gekoppelten Systemen vollständig zu verstehen und dann gezielt ausnutzen zu können, wenn Partikel in der Grössenordnung von einigen Nanometern mit hoher Effizienz gezielt auf eine Position gelenkt werden sollen. Im Folgenden wird eine Methode eingeführt, um mit niedrigem Energieaufwand einwandige Kohlenstoffnanoröhrchen, die in einem Oberflächenwachstumsprozess hergestellt worden sind, individuell in hochgradig sauberen und langzeitstabilen wässrigen Lösungen zu dispergieren. Diese Lösungen sind überaus bedeutend für eine hohe dielektrophoretische Integrationsausbeute dieser eindimensionalen Kohlenstoffnanostrukturen. Die elektrische Charakterisierung an 223 Einheiten mit niedrigem Widerstand bescheinigt die hohe Qualität der Kohlenstoffnanoröhrchenlösung, des Ausgangsmaterials und der elektrischen Kontakte. Anschliessend wird die dielektrophoretische Trennung von einzelnen metallischen Kohlenstoffnanoröhrchen aus heterogenen Lösungen nachgewiesen. Direkte elektrische Transportmessungen bestätigen ihre simultane Ablagerung zwischen Elektroden. Eine Trennungsfrequenz von 188 MHz wird von einem Oberflächenleitfähigkeitsmodell entnommen und die Abängigkeit der Trennungsfrequenz von einem eingeführten Leitfähigkeitsverhältnises untersucht. Die Vielseitigkeit der entwickelten Methoden wird durch die parallele Integration von Kohlenstoffnanoröhrchen in piezoresistive Drucksensoren unter Beweis gestellt. Kohlenstoffnanoröhrchen werden dielektrophoretisch gezielt auf den Positionen grösster Dehnung, den Membranrändern, angebracht. Höchste Sensitivität von diesen langzeitstabilen Sensoren wird bei Kohlenstoffnanoröhrchen mit kleiner Bandlücke im Bereich des kleinsten Transistorstroms gemessen. Die Sensitivität erreicht Werte von bis zu $S_0 \sim 0.25$ $\Delta R/R/$bar bei einer Auflösung von mehr als 50 mbar und einem Energieverbrauch von weniger als 40 nW. Abschliessend wird die dielektrophoretische Integration von einzel- und mehrschichtigen Graphenen, zweidimensionalen Kohlenstoffnanostrukturen, demonstriert. Die Ergebnisse dieser Arbeit ermöglichen neue Ansätze für die weitere Elektronik-, Mikro- und Nanosystemminiaturisierung. Kostengünstige lösungsbasierte Prozesse erlauben die Herstellung von Sensoren mit aktiven Elementen bis zu drei Grössenordnungen kleiner als der derzeitige Stand der Technik bei gleichzeitig erhöhter Sensitivität und signifikant reduziertem Energieverbrauch.

Acknowledgments

This thesis would never have been possible without the support and help of countless people.

First and foremost I would like to thank my advisor **Prof. Dimos Poulikakos**. Above all, I valued his trust in me and my decisions and I highly appreciated the independence I had during my doctoral studies, something which allowed me to shape this thesis according to my own interests. He was always open for new ideas and initiatives and didn't hesitate to delegate big responsibilities. His constant accessibility and feedback proved to be a great source of motivation, as well as inspiration. Also, I am deeply thankful for his active endorsement of me in the past and for the future.

The project was carried out in collaboration with the Micro and Nanosystems group of **Prof. Christofer Hierold**, hence the natural co-advisor choice. Over the past years I experienced his genuine interest in my work with very valuable input and his unconditional support for working together with members of his group. This allowed me to profit immensely from their knowledge and experience, ultimately raising the quality of the work significantly.

My project collaborator was **Thomas Helbling**. I can not say enough about how much I learned from him. He was a constant source of personnel and technical support and motivation, always open to share his latest findings and strongly encouraged my work. Especially the pressure sensor fabrication and characterization would not have been feasible without him. He provided detailed process documentation, after endless optimization, and assembled the highly automatized test rig, both foundations of his own thesis. Remarkably, and sincerely appreciated, we never got into each others way and were able to pursue our own independent objectives during our very close collaboration.

Julian Schneider deserves a lot of credit for the research reported as well. As my (only) Master's student we spent almost a year working together, including his Semester thesis. A big share of the presented results in fact originate from experiments he carried out. I truly enjoyed our exhaustive discussions, especially during phases when we didn't exactly understand observed phenomena. It was a great pleasure to see him join our group as a doctoral student and I am convinced his so precise and scrutinizing work ethic will allow him to perform an outstanding piece of research.

I had the opportunity of working intensely together with one of our guest scientists in the lab, **Vincenzo Bianco** from the University of Naples in Italy. His numerical expertise allowed an in-depth analysis of our experimentally investigated phenomena, which we wouldn't have been able to perform without him. He helped us expand our knowledge and understanding of the researched system considerably. The weekend in Naples will also always remain in fond memories.

A number of graduate and undergraduate students supported my research in the form of Bachelor's and Semester's theses. **Fabian Lütolf** conducted the first experiments on the dielectophoretic integration of graphene oxide and **Simon Maurer** followed up with pristine graphene solutions. **Anne-Charlotte Johansson** deserves a lot of credit for the dispersion of surface-synthesized carbon nanotubes, whereas **Christoph Merz** was responsible for an in depth pressure sensor membrane analysis. Finally, **Georg Kucsko** helped with AFM measurements and **Joé Bartholmé** with figure preparation. Many thanks to all of them for their committed and dedicated work.

In our group, the Laboratory of Thermodynamics of Emerging Technologies (LTNT), I immensely benefited from previous work and ever ongoing technical discussions. **Timo Schwamb** was the first to introduce me to the field of dielectrophoresis. Discussions with him and **Niklas Schirmer** were without exception very valuable and enlightening, one of the reasons I enjoyed them so much. By sharing our insights I feel that all of our projects hugely profited. Naturally all of our research overlapped and we cooperated on a daily basis. In the early stages of the project **Nicole Bieri** also contributed substantially in this respect.

Coming back to the Micro and Nanosystems group, I am deeply indebted to **Matthias Muoth** for his repeated high quality single-walled carbon nanotube synthesis and weekend palladium evaporation sessions. His experimental expertise and dedication is truly remarkable. Alumina atomic layer deposition was performed by **Kiran Chikkadi** and his profound experience with hydrofluoric etching saved me a lot of time. The carbon nanotube growth protocol goes back to **Lukas Durrer** who was very open to sharing it. I owe **Ronald Grundbacher** for his time and patience in introducing me to cleanroom fabrication and I am thankful to all the tutors of the *MEMS-Lab* for sharing their processing knowledge. Working together with all the members of this group was highly fruitful a great pleasure.

At the Particle Technology Laboratory I would like to thank **Robert Büchel** for his experimental support in the synthesis of the graphene solutions by providing necessary equipment in an unbureaucratic manner. **Antonio Tricoli**, the advisor of my first Semester's thesis, provided access to the UV-Vis-NIR spectrometer.

My longtime Luxembourgish colleague in the Nanophysics group, **Françoise Molitor**, was also very helpful by sharing cleanroom experience and recipes. It was always nice to meet her in the cleanroom and being able to chat a little bit about anything in ones mother tongue.

Philipp Rüst of the Center of Mechanics proved to be a reliable and always accessible ICP troubleshooter. His predecessor and my former mechanics teaching assistant, **Udo Lang**, was always able to give valuable input and advice on the most diverse topics. The introduction to plasma etching goes back to **Niels Quack**.

One of the most successful collaborations during my thesis originated at the Nanotech conference in Montreux where I met **Josep Puigmartí-Luis** from the new Bioanalytics group. Over the course of a year we were able to pursue some quite interesting and rewarding projects.

Ultracentrifuge access early in the project was provided by **Maja Günthert** from the Biopharmacy group.

Without all the help of the cleanroom technical staff at ETH, most of the reported work would simply not have been possible. They were always available for assistance and their commitment and patience is admiring. Namely they include in the FIRST lab **Otte Homan** who supported my chemical vapor deposition, dry etching and scanning electron microscopy activities, **Sandro Bellini** for all wetbench related activities and **Maria Leibinger** for photolithography. In the FIRST-CLA labs **Donat Scheiwiller** was always there to repair equipment which broke down under my operation, mysteriously generally towards the end of the year.

At LTNT some of my best memories are tightly bound to our "extracurricular" activities. Most notably our trips through the Balkans and to Porto, as well as enjoying common (Kebab) evenings together after work. Herewith I would like to thank all the **LTNT members** who ever shared time with me in the lab. Unfortunately there are too many of them to mention them all individually. The atmosphere in the group was ever laid-back and stimulating alike, an unbeatable combination I believe.

Slowly coming to an end, I would also like to acknowledge my regular Thursday's Luxembourgish Tech lunch company, **Jean-Claude Eischen** a.k.a. *jce*, **Georges Baatz** a.k.a. *gRgS* and **Josy Schmitz** a.k.a. *joao*. Exchanging the latest news and gossip was a weekly highlight.

I was able to run off all my work related frustrations at the weekly **TV Oerlikon** practices. The acceptance I experienced there was without doubt one of the major factors which made my thesis time so enjoyable in Zurich. I will really miss the Tuesday basketball warm-ups during the winter months, weekends in the Berghaus and training camps.

Finally, I would like to thank the most important people in my life. I can not imagine how much my girlfriend **Lynn** had to endure over the last three and a half years. Somehow I always had the feeling I needed to complain in her presence. It must have been so bad, she even started to wonder at one point why I was carrying on with my thesis at all. Nonetheless, she unconditionally supported me and my work, which included weekend shifts in the office or lab, long working hours, and writing e-mails late at night. I hope the time wasn't too bad after all... For my part, I really treasure our shared vacations during the last years. I know the future won't be easy but I hope we will be able to rise to the challenge. Thank you so much for all your support!

My **parents** also deserve a lot of credit. Even though I was financially self-sufficient during my thesis, they were an irreplaceable moral support, probably coming to Zurich more often then I made it back to Luxembourg. They always offered valuable and helpful advice when needed and their effort to understand my thesis is sincerely appreciated. I hope my public outreach was somewhat successful! My brothers **David** a.k.a. *déivi* and **Andrew** a.k.a. *N-D* also contributed their share, remembering birthdays, organizing Christmas presents and most importantly allowing for great times together and good laughs. And who knows, maybe both of them will have "real" jobs before I do!

Funding of the project was ensured by ETH Zurich research grant TH 13/05-3.

Contents

Abstract	vii
Acknowledgments	xi

1 Introduction   1
   1.1 Context 1
   1.2 Thesis Outline 3

2 Electrokinetic Framework of Dielectrophoretic Deposition Devices   5
   2.1 Introduction 6
   2.2 Numerical Model 7
      2.2.1 Geometry 7
      2.2.2 Electrical Model 8
      2.2.3 Thermal Model 9
      2.2.4 Hydrodynamic Model 10
      2.2.5 Particle Model - Dielectrophoresis (DEP) 12
      2.2.6 Variables 14
      2.2.7 Solution Strategy 14
   2.3 Electrokinetic Framework Results 15
      2.3.1 Impedance Analysis 15
      2.3.2 Electric Potential Distribution 16
      2.3.3 Temperature Field 18
      2.3.4 Fluid Velocity Field 19
   2.4 Schematic Representation of Electrokinetic Effects 25
   2.5 Experimental Confirmations 27
      2.5.1 Frequency Dependency 27
      2.5.2 DC Electroosmosis 29
   2.6 Conclusions 30

3 Aqueous Dispersion and Dielectrophoretic Assembly of Individual Surface-Synthesized Single-Walled Carbon Nanotubes   31
   3.1 Introduction 31
   3.2 Experimental Section 32

	3.3	Results	35
	3.4	Discussion and Conclusions	38

4 Selective Parallel Integration of Individual Metallic Single-Walled Carbon Nanotubes from Heterogeneous Solutions   **41**
- 4.1 Introduction 42
- 4.2 Experimental Section 42
- 4.3 Results 44
- 4.4 Discussion and Conclusions 44

5 Piezoresistive Pressure Sensors with Parallel Integration of Individual Single-Walled Carbon Nanotubes   **55**
- 5.1 Introduction 56
- 5.2 Experiments 58
- 5.3 Results 62
- 5.4 Discussion and Conclusions 63

6 High-Yield Dielectrophoretic Assembly of Two-Dimensional Graphene Nanostructures   **67**
- 6.1 Introduction 67
- 6.2 Experiments 68
- 6.3 Results 71
- 6.4 Discussion Conclusions 73

7 Dielectrophoretic Integration of Single- and Few-Layer Graphenes   **75**
- 7.1 Introduction 75
- 7.2 Experiments 77
- 7.3 Results 79
- 7.4 Discussion and Conclusions 85

8 Conclusions   **87**
- 8.1 Results Overview 87
- 8.2 Outlook 88

Bibliography   **91**

1 Introduction

1.1 Context

Fifty years ago, on December 29, 1959, Richard Feynman delivered a perspective presentation to a meeting of the American Physical Society at the California Institute of Technology [1]. What is often considered today as nanotechnology's seminal paper, *There's Plenty of Room at the Bottom*, subtitled *An Invitation to Enter a New Field of Physics*, was not immediately recognized in its significance, however. *Plenty of Room*

It was not until the invention of the scanning tunnelling microscope (STM) in the early 1980s by Gerd Binning and Heinrich Rohrer [2], enabling atomic scale resolution and allowing the manipulation of individual atoms by Donald Eigler in 1990 [3], that *Nanotechnology* truly emerged as a major field of research. Nanotechnology

These breakthroughs would not have been possible, though, without the continuous advances in miniaturization driven by the semiconductor industry. Early observations by Gordon Moore in 1965, that the number of components in integrated circuits had doubled every year since its invention, lead to his prediction that the trend would continue *for at least ten years* [4]. To date, the prediction has proven to be uncannily accurate. Moore's Law

The major driving forces behind miniaturization are cost reduction, device functionality and energy efficiency. The enablers which counter these market forces are novel materials and new process technologies. Naturally, research benefits from the outcomes and can develop, amongst others, innovative characterization tools and techniques, such as the STM for instance, which foster and encourage the miniaturization trend alike. Miniaturization

Today, semiconductor industry is mainly based on silicon (Si) as device material and complementary metal-oxide-semiconductor (CMOS) processes for fabrication, allowing the integration of complex microelectromechanical systems (MEMS) into sensors and actuators. State of the Art

Most of the approaches still heavily rely on a few ground breaking discoveries, starting with the first point-contact transistor, invented in 1947 by John Bardeen and Walter Brattain [5], shortly followed by William Shockley's work on p-n junctions and bipolar transistors in 1949 [6]. The first silicon solar cell was developed in 1954 by Chapin *et al.* [7] The most important device for advanced integrated circuits today is The Transistor

1 Introduction

the metal-oxide-semiconductor field-effect transistor (MOSFET), which was reported by Kahng and Atalla in 1960 [8].

Integrated Circuits The first integrated circuit (IC) was fabricated by Jack Kilby in 1959 [9]. Later in 1959, Robert Noyce made the first monolithic IC [10]. The processes already relied on lithography for patterning and ion implantation for semiconductor doping.

Thesis Scope Progress in miniaturization, in the semiconductor, as well as in the MEMS industry, has lead feature sizes venture into the nanometer size regime. To enable continuous miniaturization, though, novel materials and new process technologies must be researched, as some of the materials and fabrication techniques used to date will ultimately run into fundamental physical barriers. Therefore, it is proposed to move away from top-down fabrication processes toward directed assembly and use high-symmetry low-dimensional materials, while always allowing parallel assembly.

Thesis Goal In the present thesis, the dielectrophoretic integration of carbon nanostructures for parallel sensor assembly is proposed as a possible avenue to achieve this goal.

Dielectrophoresis Dielectrophoresis (DEP) is a phenomenon in which a force is exerted on a dielectric particle when it is subjected to a non-uniform electric field and was discovered by Herbert Pohl in 1951 [11]. The strength of the force depends on the medium and particles' electrical properties, on the particles' shape and size, as well as on the frequency of the electric field. Consequently, fields of a particular frequency can manipulate particles with great selectivity. The phenomenology is described in detail in a number of works, notably by Pohl [12], Jones [13], and Morgan and Green [14].

Fullerenes Carbon nanostructures, also known as fullerenes, are carbon allotropres in the form of a hollow sphere, ellipsoid, or tube. The first fullerene to be discovered, and the family's namesake, was buckminsterfullerene C_{60}, synthesized in 1985 by Robert Curl, Harold Kroto and Richard Smalley [15].

Carbon Nanotubes Single-Walled Carbon Nanotubes are cylindrical carbon structures and were simultaneously discovered by Sumio Iijima [16] and Bethune *et al.* [17] in 1993. They exhibit extraordinary strength and unique electrical properties, and are efficient thermal conductors. Their electrical conductivity can show metallic or semiconducting behavior, which make them potentially useful in many applications ranging from electronics, optics, mechanics, and other fields of materials science. Multiple reference works describe their special physical properties and potential uses [18–23].

Graphene Graphene, finally, is a one-atom-thick planar sheet of sp^2-bonded carbon atoms that are densely packed in a honeycomb crystal lattice. When Andre Geim and Kostya Novoselov managed to extract single-atom-thick crystallites (graphene) from bulk graphite in 2004, a large interest in graphene emerged, essentially because of its unique physical properties [24]. In depth reviews of graphene properties and applications are available in the meantime [25–27].

NEMS The above mentioned elements are therefore considered to help make the vision of nanoelectromechanical systems (NEMS) ultimately become a reality [28].

1.2 Thesis Outline

The outline of the thesis is given in the following.

Chapter 2 - Electrokinetic Framework of Dielectrophoretic Deposition

First, the electrokinetic framework of dielectrophoretic deposition devices is introduced. Understanding the occurrence and interaction of the different underlying effects in capacitively coupled systems, including a self-limiting integration mechanism for individual nanostructures, allows gentler particle handling without direct current throughput at an increased deposition yield.

Chapter 3 - Single-Walled Carbon Nanotube Aqueous Dispersion

A method for dispersing surface-synthesized individual, long, and large-diameter (1 − 3 nm) SWNTs in ultrapure and long-term stable surfactant-stabilized aqueous solutions by low energy input is presented in the following. Dielectrophoretic deposition and electrical characterization evidence the the high quality of the SWNT solutions, raw material, and contact interface.

Chapter 4 - Selective Dielectrophoretic Integration of Metallic SWNTs

Subsequently, the dielectrophoretic separation of individual metallic SWNTs from heterogeneous solutions is presented. Their simultaneous deposition between electrodes is confirmed by direct electric transport measurements. A threshold separation frequency of 188 MHz is extracted from a surface-conductivity model and a conductivity weighting factor introduced to elucidate the separation frequency dependence.

Chapter 5 - Parallel SWNT Based Pressure Sensor Assembly

To show the versatility of the introduced methods, the parallel integration of SWNT based piezoresistive pressure sensors is demonstrated. Carbon nanotubes are dielectrophoretically placed at the membrane edges, ultimately the positions of highest strain. Highest sensitivity of the long-term stable devices is achieved in the off-state of small band gap carbon nanotube transistors (SGS-CNFETs), reaching values as high as $S_0 \sim 0.25$ $\Delta R/R/$bar, at a resolution better than 50 mbar, and a power consumption of less than 40 nW.

1 Introduction

Chapter 6 - Few-layer Graphene Oxide Dielectrophoresis

Graphene handling is still dominated by serial mechanical exfoliation, which may well facilitate measurements in a laboratory environment but does not allow reliable larger-scale integration. Herein, the controlled, high-yield, site-selective deposition of ultrathin few-layer (three to ten) graphene oxide by dielectrophoresis between prefabricated electrodes is demonstrated.

Chapter 7 - Dielectrophoretic Graphene Integration

In the final chapter, the dielectrophoretic integration of single- and few-layered graphenes from three distinct graphene suspensions is presented. Results indicate, that the most crucial aspect for the successful thin flake deposition is the solution quality of the exfoliated graphene.

2 Electrokinetic Framework of Dielectrophoretic Deposition Devices

Parts of this chapter are published in:

B. R. Burg, V. Bianco, J. Schneider & D. Poulikakos. Electrokinetic framework of dielectrophoretic deposition devices. *Journal of Applied Physics* **107**, 124308 (2010).

Abstract

Numerical modeling and experiments are performed investigating the properties of a dielectrophoresis-based deposition device, in order to establish the electrokinetic framework required to understand the effects of applied inhomogeneous electric fields while moving particles to desired locations. By capacitively coupling electrodes to a conductive substrate, the controlled large-scale parallel dielectrophoretic assembly of nanostructures in individually accessible devices at a high integration density is accomplished. Thermal gradients in the solution, which give rise to local permittivity and conductivity changes, and velocity fields are solved by coupling electric, thermal and fluid-mechanical equations. The induced electrothermal flow causes vortices above the electrode gap, attracting particles, such as single walled carbon nanotubes (SWNTs), before they are trapped by the dielectrophoretic force and deposit across the electrodes. Long-range carbon nanotube transport is governed by hydrodynamic effects, while local trapping is dominated by dielectrophoretic forces in low concentration SWNT dispersions. Results show that by decreasing the ac frequency ac electroosmosis on the metallic electrodes occurs due to the emergence of an electric double layer, disturbing the initial flow pattern of the system. By superimposing a dc potential offset, a generated tangential electroosmotic fluid flow in the dielectric electrode gap also disrupts the electrothermal flow. Capacitive coupling is most efficient in the high frequency regime where it is the dominating impedance contribution. Understanding the occurrence and interaction of these different effects, including a self-limiting integration mechanism for individual nanostructures, allows an increased deposition yield at overall lower electric field strengths through a prudent choice of electric field parameters. The findings provide important avenues toward gentler particle handling, without direct current throughput, a relevant aspect for limiting process effects during device fabrication, all while increasing dielectrophoretic deposition efficiency in nanostructured networks.

2.1 Introduction

Solution Processing
Solution processing of electronic devices is becoming a growing field of interest due to the ability to cover large areas, the low cost of fabrication and the ease of processing. Post synthesis fabrication techniques offer attractive alternatives in contrast to the often involved high temperatures for the growth of nanoscale materials. In particular the assembly of single-walled carbon nanotubes (SWNTs) from stable dispersions is attracting a large amount of interest. The exceptional electronic properties of SWNTs make them promising building blocks for future nanoscale electronic and electromechanical devices [29].

SWNT DEP
One approach for out-of-solution guided assembly of individual SWNTs is dielectrophoresis (DEP) [11]. Dielectrophoresis allows the site-selective parallel integration of these one-dimensional nanostructures onto prefabricated electrodes by applying inhomogeneous electric fields enabling multiple (beyond two) point contacting [30].

Capacitive Coupling
By capacitively coupling one of the electrodes in the system to a conductive substrate separated by an insulating oxide layer, direct current throughput through the nanostructures is avoided during the assembly process, while minimal external contacting enables large-scale assembly [31,32]. The dielectrophoretic integration of SWNTs has been demonstrated from a wide range of solutions involving different solvents and starting raw material [32–34]. The conductivity dependent separation of SWNTs by DEP is also possible and is an inherent process characteristic [35].

Electrokinetic Framework
In order to understand, explain and quantify the underlying and occurring phenomena, an electrokinetic framework must be developed, which includes all hydrodynamic and direct particle effects in dielectrophoretic deposition devices. Qualitative, as well as quantitative insights may be thus gained to increase the deposition yield by adopting the device design and parameter settings. Previous theoretical studies have investigated direct particle effects [36], traveling wave dielectrophoresis [37], and direct electrode coupling [38], but a more general framework is still lacking.

Numerical Simulation
A comprehensive numerical study involving all effects from imposed sinusoidal voltage potentials in capacitively coupled dielectrophoretic deposition devices suited for large-scale integration will be introduced in the following. The phenomena include Joule heating from applied electric fields in aqueous electrolyte solutions, sub-

sequent emergence of permittivity and conductivity gradients in the solution leading to fluid flow, electroosmotic effects, and dielectrophoretically induced particle forces on the dispersed SWNTs, ultimately leading to their trapping between the electrodes.

In the end, experimental results confirm the conclusions drawn from the numerical simulations. SWNT solutions were prepared by removing surface-synthesized carbon nanotubes grown in a chemical vapor deposition (CVD) process by a low-energy ultrasonic pulse to produce long-term stable 1 wt % sodium dodecylbenzenesulfonate (SDBS) surfactant-stabilized aqueous solutions. The SWNT concentration in the diluted solutions is determined to lie in the range of 1 nanotube/(10 μm)^3 [34]. Dielectrophoretic deposition of the carbon nanotubes was performed on chips made up of individually accessible electrodes in the range of $1-2$ μm, according to a previously published account [34].

Experimental Confirmation

2.2 Numerical Model

2.2.1 Geometry

In the performed numerical studies, a model geometry was adopted which best matches the actual devices used in experimental studies. The chips, illustrated schematically in Figure 2.1, consisted of 20 individual positions available for dielectrophoretic deposition. The sinusoidal potential was applied to the large, connected bias electrode which extends to the finger electrodes where the dielectrophoretic deposition takes place. The counter electrodes remained floating during the deposition process to insure capacitive coupling to the grounded p-type doped silicon substrate, separated by an insulating dry thermal oxide layer [34].

Chip Template

The investigated geometry was slightly simplified for the numerical modeling, without loss of generality. Simulations were carried out on a two-dimensional cross-section of the chip, in order to limit the computational cost resulting from the large differences in length scales involved. The modeled chip domain only contains one electrode gap, located symmetrically at the center of the simulation domain. Figure 2.1 shows the dimensions of the different elements. The electrodes include the large bond pads as these are prevalent on the chip surface. The gap between the electrodes is 1 μm wide, as in the experiments and the substrate thickness is 500 μm. Above the oxide thickness of 285 nm, a solution droplet of 0.5 mm height is imposed.

Model Geometry

All simulations were carried out using an alternative current (ac), quasi-static analysis. By grounding the silicon substrate, the floating counter electrode it thus capacitively coupled to the system.

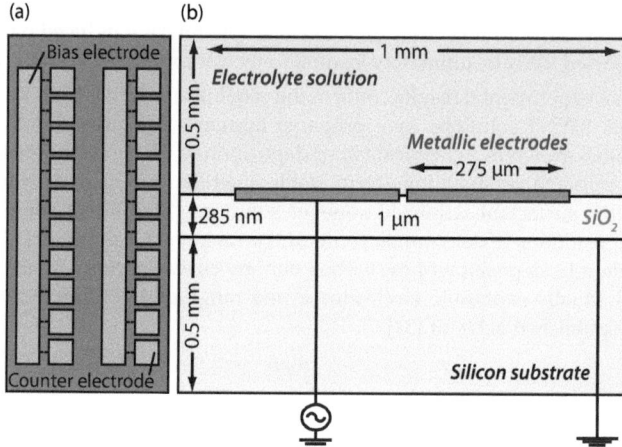

Figure 2.1: Chip and model geometry (not to scale). The simulation domain was chosen to best match actual devices used in experimental studies. All dimensions are noted. To limit computational costs, simulations were carried out on a two-dimensional cross-section of the chip. The geometry is symmetric around the electrode gap. Nonetheless the entire domain was modeled, since different boundary conditions apply on the electrodes.

2.2.2 Electrical Model

Electric Potential
For a sinusoidal potential on the bias electrode with peak voltage V_p, the potential distribution in the system of homogeneous linear dielectrics is determined by solving the equation for time harmonic quasi-statics electric current analysis:

$$-\nabla \cdot ((\sigma + i\omega\epsilon_r\epsilon_0)\nabla\tilde{\phi}) = 0, \tag{2.1}$$

with σ and ϵ_r being the conductivity and permittivity, respectively, of the dielectrics. $\tilde{\phi}$ is the complex phasor of the harmonic potential oscillating at angular frequency ω, and contains all amplitude and phase information but is independent of the time t: $\phi(\mathbf{x}, t) = \tilde{\phi}(\mathbf{x})e^{i\omega t}$. The real part of $\phi(\mathbf{x}, t)$ represents the magnitude of the applied potential. For a constant phase in the system the potential phasor remains real [14].

Electric Field
From the electroquasistatic Maxwell equations it is possible to deduce that the electric field in the system is irrotational [39, 40]:

$$\mathbf{E} = -\nabla\phi. \tag{2.2}$$

2.2 Numerical Model

The highly p-type doped silicon substrate is grounded during the electrical simulation ($V = 0$) and consequently only the fluid and silicon oxide (SiO_2) domains are solved in the numerical model. At the fluid-SiO_2 material interface the current density J, defined as $\mathbf{J} = \sigma \mathbf{E}$ for a homogeneous linear dielectric, remains constant and the following boundary condition applies: $\mathbf{n} \cdot (\mathbf{J_m} - \mathbf{J_{SiO_2}}) = 0$. Electric insulation is assumed at the model boundaries due to the large size of the simulation domain: $\mathbf{n} \cdot \mathbf{J} = 0$. Since the electric field is concentrated around the electrode gap and no significant gradients at the model edges exist, the electric insulation boundary condition is essentially equivalent to a symmetry boundary condition for the present conditions. The second electrode in the system is modeled as floating potential and is solved by applying an electric shielding boundary condition. The capacitive coupling of the second electrode is thus made possible.

Boundary Conditions

2.2.3 Thermal Model

The electric field in the system causes electric currents to arise in the electrolyte solution, which in turn give rise to Joule heating:

Joule Heating

$$Q = \left\langle \sigma |\mathbf{E}|^2 \right\rangle = \frac{1}{2} \sigma |\mathbf{E}|^2. \qquad (2.3)$$

This power generation in a very small volume induces a considerable temperature increase in the affected area. The energy balance equation allows the entailed temperature distribution, T, to be determined:

Energy Equation

$$\nabla \cdot (-k \nabla T) - Q = 0, \qquad (2.4)$$

where k is the thermal conductivity of the medium, assumed to remain constant within the limited temperature range. Viscous dissipation effects are neglected and for sufficiently high frequencies the equation can be simplified to the steady state case. Also, convection effects in the system are negligible due to the small geometry and low fluid velocities involved. No Joule heating is expected to occur in the silicon oxide dielectric [41].

All domains of the model are solved for their temperature distribution. The top and bottom side of the geometry are kept at a fixed surrounding temperature: $T = T_0$. On the edges, a symmetry boundary condition is applied: $-\mathbf{n} \cdot (-k \nabla T) = 0$. Because the edges of the modeled system are far away from the concentrated power generation near the electrode gap and temperature gradients are practically nonexistent at the edges, this boundary condition is a valid assumption.

Boundary Conditions

2.2.4 Hydrodynamic Model

Navier-Stokes Equation Electrothermal fluid flow arises from the action of an electric field on inhomogeneities in the solution medium induced by temperature gradients. The velocity of the fluid, **u**, is described by the steady-state Navier-Stokes equation, together with the mass conservation equation for an incompressible fluid:

$$\rho (\mathbf{u} \cdot \nabla) \mathbf{u} = -\nabla p + \mu \nabla^2 \mathbf{u} + \mathbf{f_e} \quad (2.5)$$
$$\nabla \mathbf{u} = 0, \quad (2.6)$$

where ρ is the density, p the pressure, and μ the dynamic viscosity of the fluid. The electric force density, also known as electrothermal force, $\mathbf{f_e}$, is added as a body force to the Navier-Stokes equation. Buoyancy forces are neglected as well as inhomogeneities in the viscosity caused by the temperature field, which are small [41].

Boundary Conditions No-slip conditions are imposed on all boundaries of the fluid domain: $u = 0$. In certain situations, however, it may be necessary to introduce a slip condition on the metallic electrodes to take the emergence of ac electroosmosis into account, especially in the low frequency regime: $u_0 = u_{aceo}$ [42]. The same applies when a dc offset is imposed between two electrodes and a slip condition on the dielectric silicon oxide substrate has to be introduced to account for dc electroosmosis: $u_0 = u_{dceo}$ [43].

Electrothermal Flow (ETF)

Electrothermal Force Local heating of the fluid gives rise to permittivity and conductivity gradients, which in turn generate a body force on the fluid. This electrothermal force written in terms of the temperature gradient is equal to [41, 44]

$$\mathbf{f_e} = \rho \mathbf{E} - \frac{1}{2}|\mathbf{E}|^2 \nabla \epsilon_m \quad (2.7)$$
$$= \frac{1}{2}\text{Re}\left[\frac{\sigma_m \epsilon_m (\alpha - \beta)}{\sigma_m + i\omega\epsilon_m}(\nabla T \cdot \mathbf{E})\mathbf{E}^* - \frac{1}{2}\epsilon_m \alpha |\mathbf{E}|^2 \nabla T\right], \quad (2.8)$$

with $\alpha = (1/\epsilon_m)(\partial \epsilon_m/\partial T) \approx -0.4\%$ K^{-1} [45] and $\beta = (1/\sigma_m)(\partial \sigma_m/\partial T) \approx 2\%$ K^{-1} [45]. $\mathbf{E}^*$ is the complex conjugate of the electric field. At low frequencies the first term in the equation, the Coulomb force, dominates and at high frequencies the second term, the dielectric force, takes over. The two forces act in different directions over a certain range of frequencies, resulting in a changing flow pattern [41, 44].

AC Electroosmosis (ACEO)

ACEO AC electroosmotic fluid flow, originating from a tangential electric field on the electrical double layer at the electrolyte-electrode interface, may emerge from sinusoidal

2.2 Numerical Model

potentials under specific frequency conditions. Alternating currents generate divergent electric fields in the planar electrode array of the dielectrophoretic deposition device and as a result, a component of the electric field lies tangential to the electrical double layer and is induced on the electrode surface. Ions in the diffuse double layer experience a force that has a time average that acts from the inner edge across the outer surface of the electrode. These charges migrate as a consequence and generate a drag flow in the fluid, which is zero directly at the slip-plane and gradually increases to the maximum velocity u_{aceo}. If the double layer characteristic thickness λ_D, also known as the Debye length, is very small compared to the characteristic length in vertical direction, the fluid can be assumed to slip at the surface. The time averaged ac electroosmotic slip velocity is derived from a generalization of the Smoluchowski formula [14, 46, 47]

Slip Velocity

$$u_{aceo} = \Lambda \frac{\epsilon_m V_p^2}{8\mu x} \frac{\Omega^2}{(1+\Omega^2)^2}, \qquad (2.9)$$

with the non-dimensional frequency Ω given by

$$\Omega = \frac{\pi \epsilon_m \omega x}{2 \sigma_m \lambda_D}. \qquad (2.10)$$

x denotes the distance in horizontal direction from the center of the electrode gap and the factor Λ is the relative capacitance of the stern layer with respect to the overall double layer capacitance and depends primarily on the conductivity of the electrolyte. The value was extrapolated from previous studies at lower solution conductivities to match the observed trend and was ultimately set to $\Lambda = 0.01$ [47].

The obtained velocity profile is highly frequency dependent and tends to zero at low and high frequency limits with a maximum velocity at $\Omega = 1$. For high frequencies, the surface charge is very low as the electric double layer does not have sufficient time to establish itself, whereas at low frequencies, the applied potential drops mainly across the double layer, thus leading to very small electric fields in the solution.

DC Electroosmosis (DCEO)

If an electric field is tangentially applied to a solid surface bathed in an electrolyte, the charges in the double layer of the electrolyte experience a force. These charges migrate as a consequence and generate a drag flow in the fluid, which is given by the Helmholtz-Smoluchowski equation for the maximum velocity just above the double layer [43]

DCEO

$$u_{dceo} = -\frac{\epsilon_m \zeta}{\mu} E_x. \qquad (2.11)$$

ζ is the zeta-potential at the slip-plane between the micelle forming surfactant solution and silicon oxide dielectric and E_x is the electric field magnitude tangential to the solid surface. If the double layer thickness is very small compared to the characteristic length in vertical direction, the fluid can be assumed to slip at the surface. This net flow over the electric double layer only occurs above dielectric materials where the tangential electric field can be established, in contrast to ac electroosmosis which only occurs above conductive materials because the potential drop across the SiO_2/electrolyte interface is negligibly small under ac conditions [42].

2.2.5 Particle Model - Dielectrophoresis (DEP)

DEP A dielectrophoretic force on a particle arises from the interaction of a non-uniform electric field on an induced particle dipole. This force does not require the particle to be charged and purely relies on its polarizability. Particles move toward regions of high electric field strength if the polarizability of the particles is greater than that of the suspending medium, otherwise they are repelled. Without a spatially varying phase, the dielectrophoretic force on a rod-shaped ellipsoid particle with its major axis parallel to the electric field lines equals [14]

$$\langle \mathbf{F}_{\text{DEP}} \rangle = \frac{1}{2} \left[\text{Re} \left(\tilde{\mathbf{p}} \cdot \nabla \right) \mathbf{E}^* \right] = \frac{1}{4} v \text{Re} \left[\tilde{\alpha} \right] \nabla \left(\mathbf{E} \cdot \mathbf{E}^* \right) \tag{2.12}$$

$$= \frac{\pi abc}{3} \epsilon_m \text{Re} \left\{ \frac{\epsilon_p^* - \epsilon_m^*}{\epsilon_m^*} \right\} \nabla |\mathbf{E}|^2, \tag{2.13}$$

with a, b, and c being the half lengths of the major ellipsoid axes, and ϵ_p^* and ϵ_m^* the complex permittivity of the particle and suspension medium, respectively. For a constant particle size in a given solution, the magnitude of the dielectrophoretic force at a specific location solely depends on the complex permittivity of the particle $\epsilon_p^* = \epsilon_p - i\frac{\sigma_p}{\omega}$ [14].

Clausius-Mossotti Factor The frequency dependency of the dielectrophoretic force is described by the Clausius-Mossotti factor

$$f_{CM} = \left\{ \frac{\epsilon_p^* - \epsilon_m^*}{\epsilon_m^*} \right\}. \tag{2.14}$$

In the low frequency limit the real part of the Clausius-Mossotti relation is only dependent on the conductivities of the particle and suspending medium, as the free charge carriers have enough time to adapt to the changing direction of the electric field. Conversely, in the high frequency limit the polarization is dominated by the permittivities of the particle and suspending medium, as the free charge carriers do not have enough time to adapt to the orientation changes of the electric field and the polarizability of the electron clouds around the atoms becomes the dominant

2.2 Numerical Model

Variable	Description	Value	Reference
k_{H_2O}	Water thermal conductivity	0.6 W (m K)$^{-1}$	[45]
k_{Si}	Silicon thermal conductivity	163 W (m K)$^{-1}$	[45]
k_{SiO_2}	Silicon dioxide thermal conductivity	1.4 W (m K)$^{-1}$	[45]
$l = (2c)$	SWNT length	$2.5 \cdot 10^{-6}$ m	
$2r(=2a=2b)$	Diameter of micellized SWNT	$3 \cdot 10^{-9}$ m	
T_0	Surrounding temperature	298 K	
V_p	Peak voltage potential	2 V	
ϵ_0	Vacuum permittivity	$8.854 \cdot 10^{-12}$ F m^{-1}	[45]
ϵ_m	Water permittivity	$80\epsilon_0$	[45]
$\sigma_{p,mSWNT}$	mSWNT conductivity	10^3 S m^{-1}	[48]
$\epsilon_{p,sSWNT}$	sSWNT permittivity	$5\epsilon_0$	[48]
ζ	Zeta potential	25 mV	[49]
λ_D	Debye length	1 nm	
μ	Dynamic viscosity of water	$0.890 \cdot 10^{-3}$ Pa s	[45]
σ_m	Solution conductivity (1 wt % SDBS)	0.250 S m^{-1}	measured
$\epsilon_{p,mSWNT}$	mSWNT permittivity	$10^3 \epsilon_0$	[48]
$\sigma_{p,sSWNT}$	sSWNT surface conductivity	2.88 S m^{-1}	[48]
$\omega = 2f$	Angular frequency	$2\pi(0.1-200)$ MHz	

Table 2.1: Variables and corresponding values used in the numerical simulations for the investigation of the electrokinetic framework in dielectrophoretic deposition devices.

mechanism for surface charging. A change in the polarizability of the particles with respect to the suspending medium at a particular crossover frequency causes a change of sign in the Clausius-Mossotti factor and consequently an inversion of the dielectrophoretic force direction. The values of the particle conductivity and permittivity for SWNTs are extracted from a previous study [48].

The particles suspended in the fluid are accelerated by the deterministic dielectrophoretic force and experience an increasing drag force, known as the Stokes drag, until they reach a terminal velocity beyond which they do not accelerate anymore. If the fluid is in motion itself, this terminal velocity also depends on the velocity of the fluid. For times much greater than a characteristic time constant $\tau_a = m/f$, where m is the mass of a particle and f is its corresponding friction factor, the particle moves at the terminal velocity [50]

Particle Velocity

$$\mathbf{u}_{SWNT} = \mathbf{u} + \frac{\mathbf{F}_{DEP}}{f}. \tag{2.15}$$

2 *Electrokinetic Framework of Dielectrophoretic Deposition Devices*

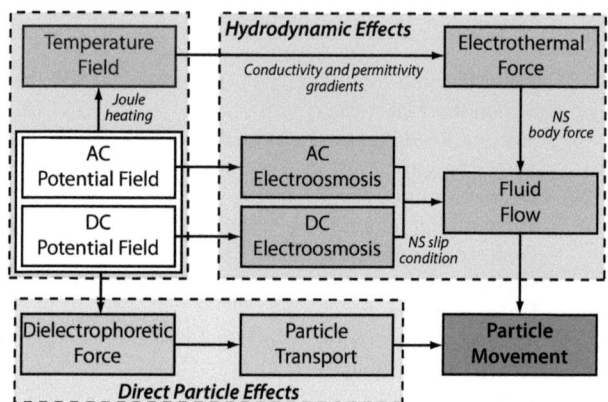

Figure 2.2: Flowchart of the model. The origin of all phenomena arises from the potential fields. Direct particle effects resulting from dielectrophoresis and hydrodynamic effects may be separated in their analysis and ultimately lead to particle movement in the system.

The friction factor for a rod-shaped particle moving at random equals [14]

$$f = \frac{3\pi\mu l}{\ln(l/r)}. \tag{2.16}$$

2.2.6 Variables

Variables Table 2.1 shows the variables and corresponding values used in the numerical simulations for the investigation of electrokinetic effects in dielectrophoretic deposition devices.

2.2.7 Solution Strategy

Sequential Solver The differential equations developed to solve for the electric field, temperature distribution and fluid dynamics of the system are coupled, but may be solved sequentially in the specific case to be studied. A schematic overview of the developed model and involved system forces is given in Figure 2.2.

The origin of all effects resides in the applied potential. This induces direct particle effects resulting from dielectrophoresis and hydrodynamic phenomena either induced by electrothermal flow arising from Joule heating or by electroosmosis. Brownian motion is neglected.

2.3 Electrokinetic Framework Results

The system was solved using a commercial software package (COMSOL MULTI-PHYSICS v3.5a). A sensitivity analysis was carried out to ensure grid independence and a structured mesh with 107 000 elements was adapted in the finite element method (FEM) simulation, with a higher density of grid elements in the vicinity of the electrode gap.

COMSOL

2.3 Electrokinetic Framework Results

2.3.1 Impedance Analysis

Impedance measurements of capacitively coupled dielectrophoretic deposition devices are important to understand the electrical contributions of all system elements during dielectrophoresis. The impedance was measured between the bias electrode (BE) and counter electrode (CE) of the chip, using an impedance analyzer (Solartron SI 1260) and neglecting contributions from the leads. As shown in Figure 2.3, an equivalent circuit of two capacitors connected from each electrode to the back gate (BG) through the oxide layer can be adopted for a bare chip without electrolyte solution on top. The capacitance of the system will be dominated by the capacity between the counter electrode and back gate C_{CE-BG}, as the counter electrode is about 10 times smaller than the bias electrode. From measurements, an overall back gate capacitance of $C_{BG} \approx 10$ pF is extracted.

Back Gate Impedance

By adding electrolyte solution on top of the chip, an additional fluid resistance and capacitance, as well as a constant phase element (CPE) must be introduced to the equivalent circuit. The constant phase element is used to model the double layer between the electrodes and the electrolyte and has an impedance of $Z_{CPE} = \frac{1}{C_{dl}(i\omega)^p}$, where p characterizes the imperfect capacitance of the double layer [14]. By numerically fitting the model to the impedance measurements, a double layer capacitance of $C_{dl} \approx 50$nF with an exponent of $p \approx 0.8$ is found. The capacity in the electrolyte solution is approximated from impedance measurements made with de-ionized (DI) water. Figure 2.3 illustrates the good agreement between the model fit and the obtained measurements in 1 wt % SDBS aqueous solutions.

Double Layer Impedance

In figure 2.3, three regimes are distinguished in the impedance measurements for chips covered by electrolyte solution. At frequencies below 50 kHz, the double layer capacitance dominates the impedance. At frequencies between 50 kHz and 1 MHz, the impedance is governed by the conductance of the electrolyte, thus barely dependent on the frequency. At frequencies above 1 MHz, the impedance is predominantly influenced by the back gate capacitance. In this frequency range the capacitive electrode coupling for parallel dielectrophoretic deposition is the most efficient.

3 Regions

It can be deduced from the impedance analysis, that ac electroosmosis affects

Debye Length

2 Electrokinetic Framework of Dielectrophoretic Deposition Devices

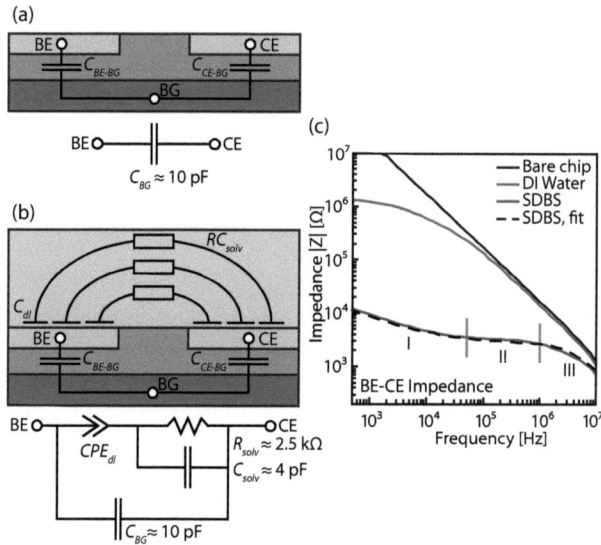

Figure 2.3: Impedance spectroscopy of the dielectrophoretic deposition device. (a) Bare chip modeled by the back gate capacitance C_{BG}. (b) For an electrolyte solution on the chip, a (non-ideal) double layer capacitance is integrated into the equivalent circuit using a constant phase element. (c) Impedance measurements between the bias and counter electrode. The dashed line is a fit to the electrolyte-chip impedance using the equivalent circuit. The three regimes of the system impedance are clearly distinguished.

fluid flow mainly below 100 kHz, where a stable double layer formation occurs. By assuming that the measured double layer capacity is governed by the small counter electrode and by treating the double layer as an ideal capacitor, the Debye length can be roughly estimated:

$$\lambda_D = \epsilon_m \epsilon_0 \frac{A_{CE}}{C_{dl}} \approx 1 \text{ nm}. \tag{2.17}$$

This is in good agreement with estimations for monovalent, symmetrical and fully dissociated salt in aqueous solutions [43].

2.3.2 Electric Potential Distribution

1 MHz Solving the time harmonic quasi-statics potential distribution in the dielectrophoretic deposition device gives insights into the capacitive electrode coupling embedded in

2.3 Electrokinetic Framework Results

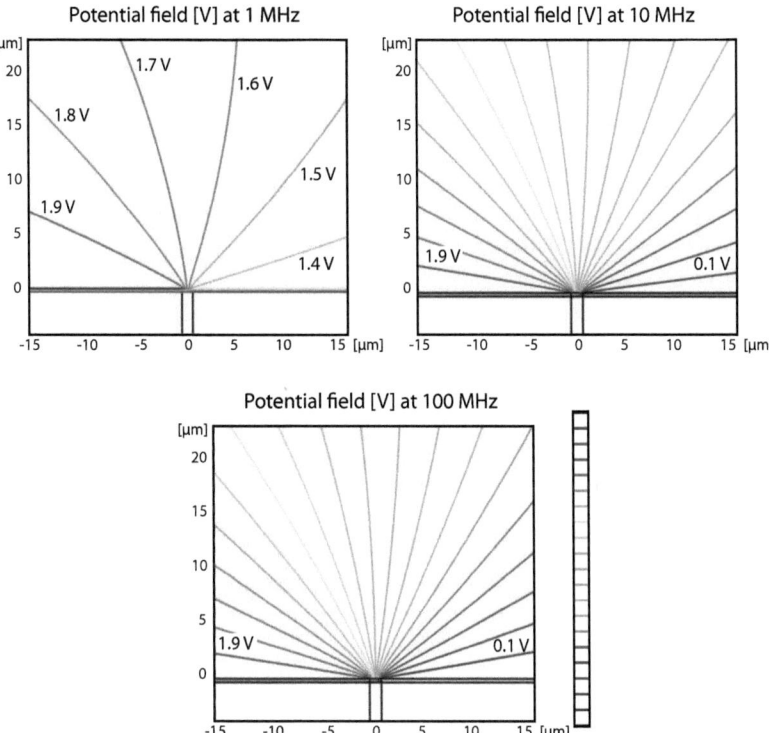

Figure 2.4: Electric potential distribution at 1 MHz, 10 MHz, and 100 MHz. Below 10 MHz the solution conductivity is still dominant and the capacitive coupling limited. Therefore the counter electrode does not take up the potential of the back gate and a limited potential drop across the electrodes is observed. A reduced fluid flow will ultimately be the consequence. At 10 MHz and above, the potential field fully develops with the strongest potential drop close to the bias electrode, leading to non-symmetric temperature and velocity fields. Almost ideal capacitive coupling is assured under these conditions.

the electrolyte solution. Figure 2.4 shows the potential distribution for 3 distinct frequencies at 0° phase. At 1 MHz, the potential applied on the left bias electrode extends deep into the solution due to the comparatively high solution conductivity. The electrolyte conductivity accounts for the dominating impedance contribution at the investigated frequency, while the back gate capacity only has a limited influence, as can be inferred from transition of region II to region III in the impedance analysis. A maximum potential difference of 0.7 V between the bias and counter electrode is achieved for an applied potential of 2 V. The electric field between both electrodes is thus limited and consequently influences the dielectrophoretic force magnitude, dependent on the square of the electric field gradient. Small potential differences also reduce the current flow in the solution, which leads to less Joule heating and ultimately results in a decreased electrothermal flow.

10-100 MHz Above 10 MHz, the influence of the back gate capacity dominates the system impedance, clearly corresponding to region III of the impedance measurements. The higher frequencies allow the applied potentials to develop to their full extent between the electrodes. Therefore little difference between 10 MHz and 100 MHz in the potential field is seen. The counter electrodes take up the ground potential of the silicon back gate, demonstrating the effectiveness of the capacitive coupling above 10 MHz and static analysis would be possible to be carried out in this frequency range..

Non-Symmetries An additional feature of the capacitive coupling is the non-symmetric potential and electric field distribution across the electrode gap. The electric potential drops considerably stronger in the vicinity of the bias electrode than counter electrode. This inhomogeneity in the electric field distribution will lead to non-symmetric induced temperature and velocity fields.

2.3.3 Temperature Field

Temperature Gradients Inserting the resulting Joule heating from the applied electric potential as external heat source into the energy equation, the temperature field in the system is obtained. As seen in Figure 2.5, the resulting temperature differences are not very large on an absolute scale. Due to the small electrode gap dimension, however, the temperature gradients are significant, especially in the gap vicinity. These temperature gradients lead to conductivity and permittivity gradients in the surfactant stabilized solution and cause electrothermal fluid flow.

2.3 Electrokinetic Framework Results

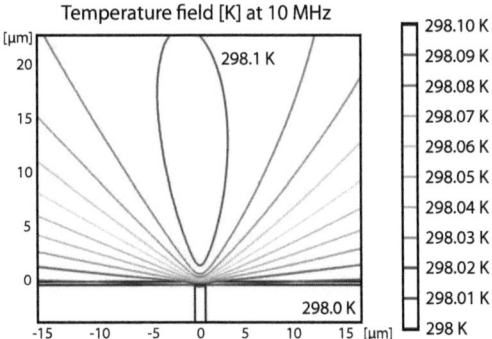

Figure 2.5: Temperature field at 10 MHz. Despite small temperature variations on an absolute scale, large gradients at the electrode gap vicinity are observed due to the small system dimensions. This will lead to conductivity and permittivity gradients, resulting in electrothermal fluid flow.

2.3.4 Fluid Velocity Field

Electrothermal Flow

After having determined the potential and temperature fields in the system, the fluid velocity field in the dielectrophoretic deposition device can be worked out. The velocity field is dominated by electrothermal forces at high frequencies, where the electric double layer is not able to establish itself. Region III of the impedance measurements corresponds to this domain and is considered to take place above 1 MHz. Two regimes occur in electrothermal flow, as described by Equation 2.8. At lower frequencies the so-called Coulomb force dominates, whereas at higher frequencies dielectric forces take over. A crossover frequency separates both regimes, where an intermediate flow pattern prevails. From Equation 2.8, this electrothermal flow crossover frequency is given by [41]:

$$f_c \approx \frac{1}{2\pi} \frac{\sigma}{\epsilon} \left| 2\frac{\beta}{\alpha} \right|^{0.5} \approx 178 \text{ MHz} \qquad (2.18)$$

ETF Crossover

Figure 2.6 depicts the different flow behavior patterns. Below the electrothermal flow crossover frequency, at 50 MHz, fluid is attracted toward the electrode gap from above and moved away tangentially to the electrode surface. Increased fluid movement close to the counter electrode stems from the non-symmetric electric potential distribution between the electrodes due to the capacitive coupling of the floating

Coulomb Forces

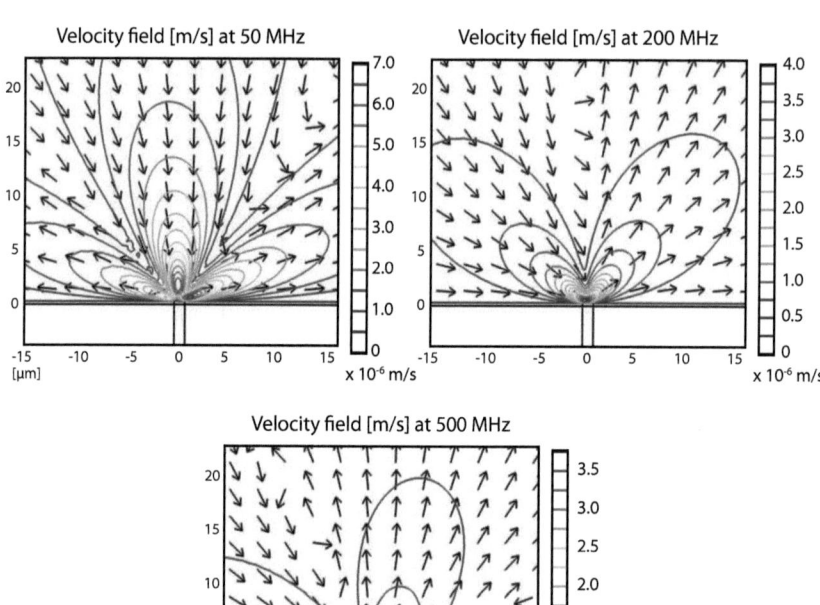

Figure 2.6: Electrothermal fluid flow velocity field at 50 MHz, 200 MHz, and 500 MHz. The solid lines are a counter plot of the velocity field and represent isotachs, while the arrows illustrate the orientation of the fluid flow. Below the electrothermal flow crossover frequency, at 50 MHz, the Coulomb force contribution in the electrothermal flow dominates. The fluid flow is described by two vortices, transporting solute from above toward the electrode gap. In the range of the ETF crossover frequency, such as 200 MHz, the Coulomb and dielectric forces in the system compensate each other. A single vortex counterclockwise pattern is observed. Above the ETF crossover frequency, at 500 MHz, the dielectric force contribution to the ETF dominates. The fluid flow changes its orientation and is described by two vortices transporting solute vertically away from the electrode gap.

2.3 Electrokinetic Framework Results

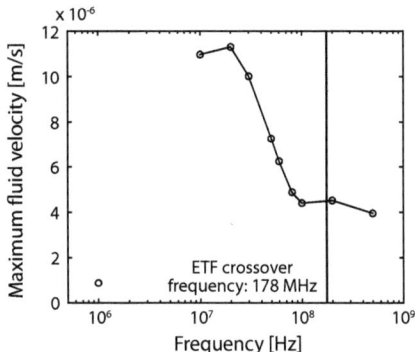

Figure 2.7: Maximum electrothermal fluid flow velocity at different frequencies. A gradual decrease is observed from the maximum around 10 − 20 MHz to the ETF crossover frequency where the maximum velocities remain constant up to reasonably high frequencies of 500 MHz for dielectrophoretic deposition devices. Reduced electrical field strengths explain the significantly lower flow velocities at 1 MHz.

counter electrode to the grounded silicon substrate. The fluid flow essentially describes two vortices above the electrodes, transporting solute from above toward the electrode gap.

At frequencies significantly higher than the electrothermal flow crossover frequency, for example 500 MHz, the orientation of the flow changes. Fluid is attracted toward the electrode gap tangentially along the electrodes and pushed upwards, which can be described by two vortices, transporting solute vertically away from the electrode gap. Again, non-symmetries in the flow pattern are due to the inhomogeneous potential distribution.

Dielectric Forces

Intermittent frequencies, such as 200 MHz, where the Coulomb and dielectric forces compensate each other, experience a single vortex counterclockwise flow profile. The magnitude of the fluid velocity remains in the same order of magnitude. Figure 2.7 compares the maximum fluid velocities in the investigated system for different frequencies. The highest fluid velocities are obtained for frequencies in the range of 10 − 20 MHz. A gradual decrease of the fluid velocities to the electrothermal flow crossover frequency is observed, where they remain constant up to reasonably high frequencies of 500 MHz for dielectrophoretic deposition devices. Significantly lower velocities obtained at 1 MHz are due to the reduced electrical field strength.

Crossover Pattern

2 Electrokinetic Framework of Dielectrophoretic Deposition Devices

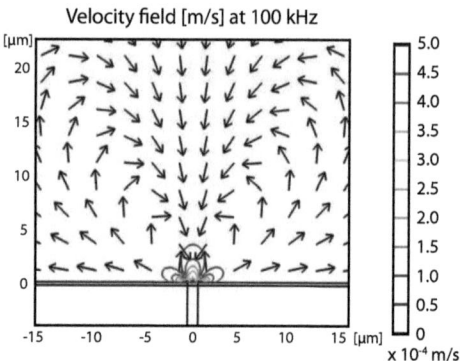

Figure 2.8: Fluid flow induced by ac electroosmosis at 100 kHz. Electrothermal effects are negligible at these frequencies, even though the flow patterns are similar. Maximum velocities are achieved close to the electrode gap just above the electrodes and are over one order of magnitude larger than the maximal electrothermal forces present at high frequencies.

AC Electroosmosis

Electrode Slip   In the lower frequency ranges, where the electric double layer has sufficient time to build up, ac electroosmosis arises. The flow is characterized by a slip velocity on the metallic electrodes which drags fluid away from the electrode gap. Consequently, solute is attracted from above. Highest velocities are reached very close to the electrode gap, just above the electrodes. As Figure 2.8 shows, ac electroosmotic flow displays a similar flow pattern to electrothermal flow for the investigated configuration, even though the origin is completely different. AC electroosmosis is due to surface effects and electrothermal flow is due to body forces in the fluid. This also explains the up to one order of magnitude higher velocities observed when ac electroosmosis occurs. The electrothermal body force has a negligible contribution in this situation. AC electroosmotic flow is strongest when the electric double layer can fully develop, corresponding to region I, below 50 kHz according to the impedance analysis presented in Figure 2.3.

DC Electroosmosis

Gap Slip   When a tangential static electric potential is superimposed on the system, for example by applying an additional voltage difference between the two bias electrodes in Figure 2.1, a drag force on the solution above dielectric surfaces of the chip is

2.3 Electrokinetic Framework Results

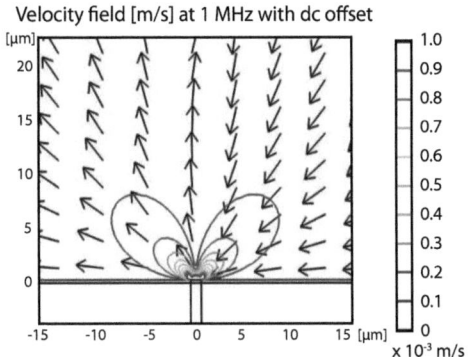

Figure 2.9: Fluid flow induced by dc electroosmosis at 1 MHz. An applied dc offset causes a fluid flow tangential dielectric surfaces. This slip flow in the electrode gap generates a vortex rotating in clockwise direction above the chip, disrupting the electrothermal flow caused by ac potentials.

created. In the fluid area located between both electrodes, charges above the electric double layer migrate along the imposed tangential electric field.

In Figure 2.9 the flow behavior for an additionally generated dc offset is illustrated. At 1 MHz, the experienced dc electroosmotic flow exceeds the electrothermal flow and is the dominating contribution in the flow pattern. The slip flow above the silicon oxide layer in the electrode gap along the electric field causes the solution to rotate in clockwise direction above the chip geometry. Uncertainties in the zeta-potential of the electric double layer require the obtained velocity values to be treated with caution.

Dielectrophoresis

Single-walled carbon nanotubes dispersed in solution experience a dielectrophoretic force when they are subjected to an inhomogeneous electric field. Both, metallic and semiconducting SWNTs exist and must be treated separately in the analysis. In order to estimate the region of influence of the dielectrophoretic force, it was investigated where the induced terminal particle velocity exceeds the maximum velocity of the electrokinetic fluid flow. mSWNTs and sSWNTs

From Figure 2.10, it is apparent that the dielectrophoretic force is highly location dependent due to the influence of the electric field gradients and that the respective DEP force magnitudes for metallic SWNTs and semiconducting SWNTs differ con- Force Range

2 Electrokinetic Framework of Dielectrophoretic Deposition Devices

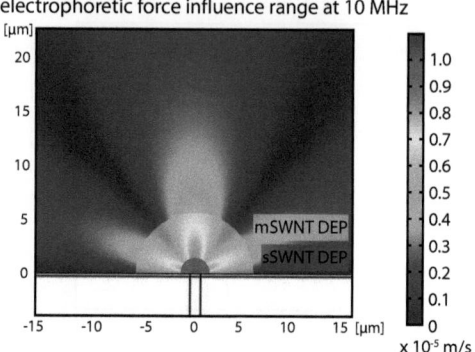

Figure 2.10: Regions of dielectrophoretic force influence where the terminal particle velocity exceeds the maximum velocity of the electrokinetic flow. The DEP force magnitude is highly location dependent and differs considerably for metallic and semiconducting SWNTs. The frequency is 10 MHz.

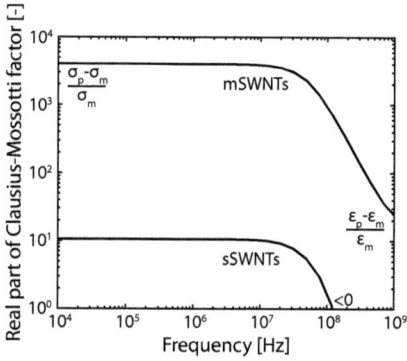

Figure 2.11: Frequency dependency of the real part of the Clausius-Mossotti factor for SWNTs. At low frequencies the Clausius-Mossotti relation depends solely on the conductivities of the particle and suspending medium, whereas at high frequencies the polarization is dominated by the permittivities of the particle and suspending medium. For semiconducting SWNTs, the Clausius-Mossotti becomes negative in the high frequency limit and the carbon nanotubes are repelled from the electrode gap.

siderably. Because of the much higher conductivities and permittivities for metallic SWNTs, the DEP force ranges much deeper into the solution. The magnitude of the DEP force is also dependent on the applied electric potential frequencies, even when perfect capacitive coupling is ensured. The reason resides in the frequency dependency of the Clausius-Mossotti factor, as can be inferred from Equation 2.14. At increasing frequencies the permittivities of the particle and suspending medium start to take over and the Clausius-Mossotti factor changes its value accordingly, as shown in Figure 2.11. If the Clausius-Mossotti factor becomes negative, the dielectrophoretic force changes orientation and the particles are repelled from sites of high electrical field strength. DEP Crossover

2.4 Schematic Representation of Electrokinetic Effects

In an effort to describe all electrokinetic effects of dielectrophoretic deposition devices for single-walled carbon nanotubes in a single representation, Figure 2.12 was constructed. At characteristic frequencies, the schematics show the interaction of the different electrokinetic components, including electrothermal flow, dielectrophoresis, ac electroosmosis and dc electroosmosis. The arrow thickness symbolizes the corresponding relative force strength. Flow Schematics

At 100 kHz, corresponding to region I from the impedance measurements in Figure 2.4, the electric double layer has sufficient time to build up and ac electroosmosis above the metallic electrodes is the dominating contribution in the fluid flow. Electrothermal effects are vanishingly small. The insufficient capacitive coupling of the counter electrodes causes a limited electric field to be created between the electrodes and therefore the dielectrophoretic forces are also small. Region I

Above 1 MHz, no stable electric double layer forms anymore, consequently ac electroosmosis disappears and electrothermal flow in the system arises. For higher frequencies the electrothermal flow rises in magnitude due to the increasing electric fields strength between the electrodes, resulting from the more efficient capacitive electrode coupling. An intermittent flow regime is observed near the electrothermal flow crossover frequency at 200 MHz, where the dielectric force effects start to take over from the Coulomb force and reverse the orientation of the vortices ultimately. Region III

The dielectrophoretic force magnitude is strongly dependent on the electric field gradient in the electrode gap and the same amplifying behavior applies between 1 and 10 MHz as for the electrothermal flow. Further, at high frequencies the polarizability of the particles in the medium is dominated by their permittivities as inferred from Figure 2.11. Above 200 MHz, the dielectrophoretic force on semiconducting DEP

2 Electrokinetic Framework of Dielectrophoretic Deposition Devices

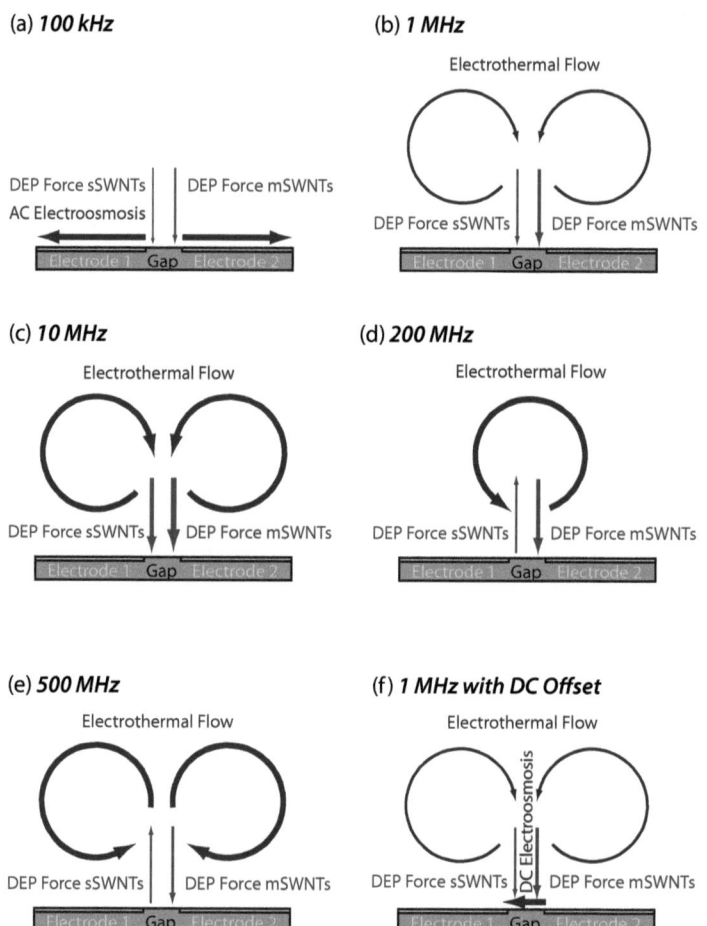

Figure 2.12: Schematic representation of the electrokinetic effects in the investigated system. The interaction of electrothermal flow, dielectrophoresis on metallic and semiconducting SWNTs, ac electroosmosis and dc electroosmosis is shown at characteristic frequencies. The compilation illustrates the electrokinetic framework of capacitively coupled dielectrophoretic deposition devices for single-walled carbon nanotubes.

SWNTs becomes negative and reduces significantly for metallic SWNTs.

In the case of an additional horizontal dc offset in the system, a slip velocity above the electrode gap is introduced. This may disrupt the flow pattern, as well as the dielectrophoretic trapping efficiency. DC Offset

Qualitatively, long-range carbon nanotube transport is governed by hydrodynamic effects, while local trapping is dominated by dielectrophoretic forces in low concentration SWNT dispersions. Fluid is attracted toward the electrode gap by the former, before the SWNTs are dielectrophoretically aligned between the electrodes by the latter, once they are within the range of the dielectrophoretic force. Qualitative Description

Fundamental time scales of the dielectrophoretic deposition process are seconds, which is explained by the length scales of the system in the order of micrometers and the fluid velocities of the electrothermal flow in the range of micrometers per second. Also, distances between individual SWNTs in the dilute dispersed solution are estimated to be within the micrometer range. Therefore voltage potentials were applied for a duration of 60 s in the experiments to ensure successful carbon nanotube deposition. Time Scales

2.5 Experimental Confirmations

2.5.1 Frequency Dependency

In an effort to experimentally verify elements of the predicted phenomena, a frequency dependent dielectrophoretic deposition yield study at constant electrical potentials was conducted at a peak potential of $V_p = 3$ V for a duration of 1 min. Parameters

At 100 kHz no deposition is observed, which is explained by the practically absent capacitive coupling of the counter electrodes to the chip substrate. Very small electric field gradients between the bias and counter electrodes result in a very weak dielectrophoretic force. Additionally, the ac electroosmotic drag force over the electrodes possibly prevents the pinning of SWNTs onto the electrodes. Region I

By increasing the frequency into the MHz-regime the amplifying capacitive coupling increases the dielectrophoretic force and the deposition of SWNTs is observed. Also, electrothermal flow develops, transporting the carbon nanotubes from the dilute solutions into the vicinity of the electrodes where they may be dielectrophoretically trapped. Region II

Maximum transport velocities and most efficient capacitive coupling found in the simulations and impedance measurements around 10 MHz is clearly confirmed by the experimental results. At 100 MHz the deposition yield drops again, as the threshold frequency of positive to negative DEP for semiconducting SWNTs is approached and about two-thirds of all SWNTs are gradually excluded from deposition. Further, Region III

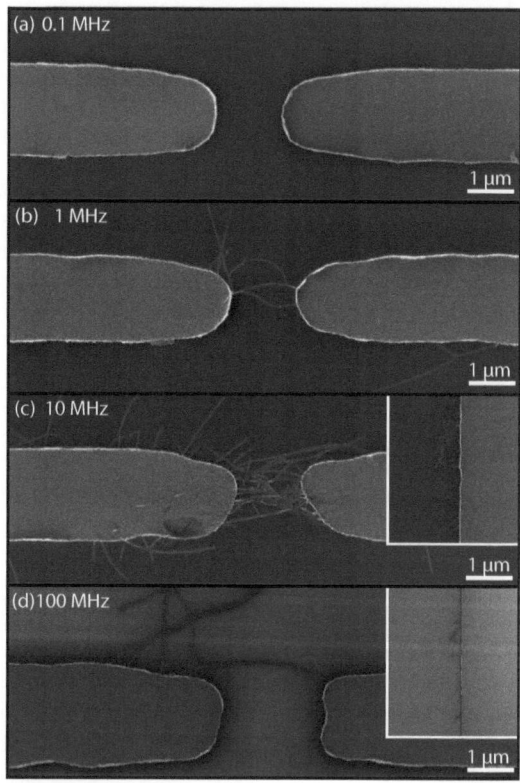

Figure 2.13: Frequency dependent DEP deposition yield studies at constant potentials. (a) No deposition is observed at 100 kHz due to the practically absent capacitive coupling of the counter electrodes to the chip substrate. (b) Limited deposition is observed at 1 MHz. (c) Strongest deposition is observed at 10 MHz, where fluid and particle transport is predicted to be the highest. (d) Reduced deposition at 100 MHz is explained by the decrease in fluid flow and gradual exclusion of semiconducting SWNTs in the deposition process resulting from their transition to negative DEP.

2.5 Experimental Confirmations

the Coulomb term in the electrothermal force reduces its influence, thus diminishing the long range drag forces in the solution toward the electrode gap. Losses in the electrical connections at these high frequencies may play a role as well. The insets in Figure 2.13 at 10 MHz and 100 MHz show the deposition of SWNTs at electrode edges, which arises from increasing field gradients caused by the small propagation distance of the electric potential into the electrolyte solution.

To achieve the dielectrophoretic deposition of individual SWNTs, the potential has to be individually adapted for every frequency applied. A self-limiting assembly mechanism exists in a narrow potential window, where the potential between bridged electrodes evens out after deposition, so that the capacitively induced electric field is significantly reduced and no additional carbon nanotubes are deposited. The potential equilibration between the electrodes is established by the inherent material conductivity of metallic SWNTs or the surface conductivity of semiconducting SWNTs induced from the surfactant-stabilized solution [48]. When the applied potential or solution concentration is too high, the self-limiting mechanism does not hold anymore and multiple carbon nanotubes are deposited. This is because the remaining electric field is still strong enough to attract additional SWNTs, as seen in Figure 2.13. The direct coupling of electrodes does not allow a self-limiting mechanism to occur, since the potential difference between the electrodes always remains the same. Further, direct coupling induces current throughput along the deposited SWNTs during the integration process, which may harm the structures.

Self-Limiting Mechanism

Direct Coupling

2.5.2 DC Electroosmosis

The influence of a 1 V dc offset at 1 MHz was additionally investigated on the dielectrophoretic deposition yield. Dielectrophoresis on chips with an identical root-mean-square potential V_{rms} was carried out and the total number of deposited SWNTs and bridging SWNTs between the electrodes counted. Figure 2.14 shows the processed results. The superimposed dc offset yields a statistically relevant lower deposition yield over their entire potential range. Two effects explain the observation. First, the dc potential causes an electroosmotic flow laterally above the electrode gap thus lowering the deposition probability of the carbon nanotubes by dragging them away from the deposition site. A secondary effect originates from the reduced torque on the induced dipole of the SWNTs due to the reduced ac electric field when a dc offset is added and the root-mean-square potential maintained. An inferior alignment in the electric field is the result, which lowers the effective dipole moment of the SWNTs and consequently reduces the dielectrophoretic force [14].

Reduced Deposition

2 *Electrokinetic Framework of Dielectrophoretic Deposition Devices*

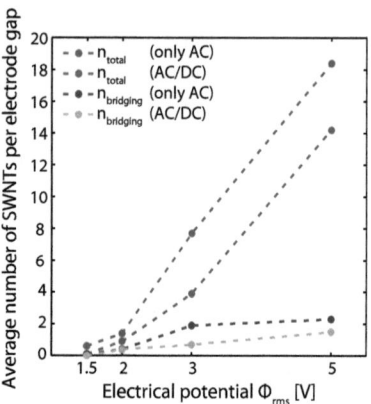

Figure 2.14: DC offset influence on dielectrophoretic deposition yield. A significant reduction for both bridging and overall deposited SWNTs is observed.

2.6 Conclusions

3 Regions Three regions, distinguished by the electrical impedance measurements, are important to describe the electrokinetic framework of capacitively coupled dielectrophoretic deposition devices. The first is the frequency range where the electric double layer of the electrolyte arises, the second where the conductivity of the electrolyte takes over as dominating contribution in the system impedance, and the third where the capacitive coupling of the electrodes to the chip substrate becomes dominating. Naturally, these frequency dependent regions depend on the electrolyte solution and chip configurations used and have to be adapted accordingly. The electrothermal and dielectrophoretic forces in the system make up additional frequency dependent

System Configuration behavior, which is also relevant in the analysis. A thorough determination of the configuration to be researched and the involved properties must therefore be made before to applying the herein developed general framework to other applications and systems, such as the dielectrophoretic integration of two-dimensional graphene sheets [51, 52]. In general, however, to achieve a high dielectrophoretic integration yield it is recommended to carry out the dielectrophoretic deposition at frequency ranges where the capacitive coupling is the dominating impedance contribution, hydrodynamic mixing effects are strongest to ensure efficient particle transport to the deposition site and where the Clausius-Mossotti factor of the dielectrophoretic force is largest.

3 Aqueous Dispersion and Dielectrophoretic Assembly of Individual Surface-Synthesized Single-Walled Carbon Nanotubes

Parts of this chapter are published in:

B. R. Burg, J. Schneider, M. Muoth, L. Durrer, T. Helbling, N. C. Schirmer, T. Schwamb, C. Hierold & D. Poulikakos. Aqueous dispersion and dielectrophoretic assembly of individual surface-synthesized single-walled carbon nanotubes. *Langmuir* **25**, 7778-7782 (2009).
http://pubs.acs.org/articlesonrequest/AOR-iF9x7SvDZSgPXdCyVf9B

Abstract

The successful dispersion and large-scale parallel assembly of individual surface-synthesized large-diameter (1 − 3 nm) single-walled carbon nanotubes (SWNTs), grown by chemical vapor deposition (CVD), is demonstrated. SWNTs are removed from the growth substrate by a short, low-energy ultrasonic pulse to produce ultrapure long-term stable surfactant-stabilized solutions. Subsequent dielectrophoretic deposition bridges individual, straight, and long SWNTs between two electrodes. Electrical characterization on 223 low-resistance devices ($R_{average} \approx 200$ kΩ) evidences the high quality of the SWNT raw material, prepared solution, and contact interface. The research reported herein provides an important framework for the large-scale industrial integration of carbon nanotube-based devices, sensors, and applications.

3.1 Introduction

The availability of individually dispersed single-walled carbon nanotube (SWNT) solutions [53–55] is of prime importance for the progress toward monodisperse

SWNT Dispersion

carbon nanotube samples [56]. Immediately after the discovery of carbon nanotubes [16,17,57], these one-dimensional structures have been the subject of intense research, both in fundamental physical studies [22] as well as in the development of nanoelectronic devices, with nanotubes as the active transducer elements [23]. The defining characteristics of SWNTs are determined by their chirality [18], affecting the electric and optical properties of the material. Since all potential technologies involving SWNTs require predictable and uniform performance, samples of SWNTs with well-defined properties under conditions allowing easy further processing must be made available. Typically, in SWNT solution processing, raw nanotube products from a bulk synthesis high-pressure CO (HiPCO) reactor process are ultrasonically treated for 10 min at elevated energies before being centrifuged at very high accelerations to remove remaining bundles [53]. In the present paper, a method for dispersing surface-synthesized individual, long, and large-diameter (1 − 3 nm) SWNTs in ultrapure and long-term stable surfactant-stabilized aqueous solutions by low energy input is presented. The process is an important contribution in the research of selective SWNT growth for subsequent dispersion [58] and chirality dependent chemical functionalization of carbon nanotubes [59].

DEP Assembly
The quality of the prepared solutions and viability of the process are assessed by the large-scale parallel dielectrophoretic integration and subsequent electrical characterization of individual SWNT-based devices. Dielectrophoresis (DEP) allows the selective deposition of micro- and nanoscale objects in nonuniform electric fields [11]. Particles move toward regions of high electric field strength if the polarizability of the particles is greater than that of the suspending medium, otherwise they are repelled [12–14]. Results on the alignment of carbon nanotubes [60], their individual, suspended, 4-point contacted deposition [30], the separation of metallic from semiconducting SWNTs [35], and their large-scale assembly [32] have been reported employing the method. Devices for physical property characterization have further been created with this technique [61,62], and more recently the deposition of two-dimensional graphene-based nanostructures was achieved [34]. Most importantly, large-scale parallel dielectrophoretic deposition enables the fabrication scale-up of a large number of devices on a single prefabricated chip, which can be then integrated into a wide range of micro- and nanoelectro-mechanical systems (MEMS and NEMS) and transducer applications, such as pressure, gas, flow, or temperature sensors.

3.2 Experimental Section

SWNT Growth
In the present study, single-walled carbon nanotubes are grown in an iron (Fe)-catalyst-based chemical vapor deposition (CVD) process [63]. In short, 1 mL of iron loaded horse spleen Ferritin (Fluka, 50 − 150 mg/mL) is added to 100 mL deionized

3.2 Experimental Section

water (DIW) and dialyzed for 24 h against it. The solution is then centrifuged at 12 000g for 15 min, and the upper half is decanted to reduce protein aggregates. The Ferritin solution is adsorbed on 7×7 mm^2 silicon chips, covered by a 1.5 µm wet oxide layer, by dipping them into the solution for 1 min. After rinsing the chips in DIW, the protein shells are oxidized and the remaining ferrihydrite particles are calcined at temperatures above 800 °C for 1 min. A PEO 603-PLC-300C LPCVD oven (ATV Technology) is used for the SWNT growth. In a first step, the Fe$_2$O$_3$ nanoparticles are reduced in H$_2$ at 850 °C at a pressure of 200 mbar for 10 min. Subsequent tube growth occurs at the same temperature under a H$_2$/CH$_4$ atmosphere with a pressure of 50/150 mbar for 30 min. Batch processing permits nanotube growth on a large number of chips, generally 25 per run. SWNT diameters from this process lie in the range of $1 - 3$ nm, revealing individual lengths of over 10 µm, with on average 2 µm and very few kinks and bends [63].

Since the generated electric field between the electrodes required for the dielectrophoretic deposition is only effective very locally at the length scale of the gap size, enough SWNTs must be present in the vicinity of the electrodes to facilitate a successful deposition. Consequently, the prepared solutions must be sufficiently concentrated. This is achieved by successively dispersing the surface-synthesized SWNTs from the typically 25 chips they were grown on into a 1 mL, 1 wt % sodium dodecylbenzenesulfonate (SDBS) DIW solution. The carbon nanotubes are removed from the SiO$_2$ growth support surface by a short, low-energy 1 s ultrasound pulse from a horn sonicator (Sonics Vibracell VCX 130, 130 W, 20 kHz). The ultrasound probe, with a tip diameter of 6 mm, is closely placed 1-2 mm over the substrate. An illustration of the applied aqueous dispersion process is depicted in Figure 3.1. SWNT Dispersion

The dispersed SWNT material is characterized by electrical measurements on prefabricated electrodes after the directed assembly. The 50 nm thick electrodes are fabricated by optical lithography, palladium (Pd) sputtering and lift-off, with gaps of $1 - 2$ µm created between them. On one 2×4 mm^2 chip, 20 individually accessible electrode gaps are available. An insulating dry thermal gate oxide layer ($t_{ox} = 285$ nm) allows the capacitive coupling between the highly p-type doped silicon substrate and the Pd electrodes for large-scale assembly [31]. This permits minimal external contacting for depositing SWNTs over multiple electrode gaps and avoids direct current throughput in the SWNTs during the deposition process. The capacity of the connected system is in the range of $C \approx 10$ pF. An aggressive oxygen plasma cleaning procedure after the electrode fabrication process results in very clean and flat sample surfaces. Chip Fabrication

A 3 µL droplet of synthesized aqueous SWNT solution is dispensed on the chip for the dielectrophoretic nanotube assembly, and a sinusoidal potential difference is applied to the contacted electrodes (typically 10 MHz, 3 V$_p$) through an sinusoidal function generator (LeCroy LW420B). The applied potential is constantly monitored SWNT DEP

3 Aqueous Dispersion and Dielectrophoretic Assembly of Individual SWNTs

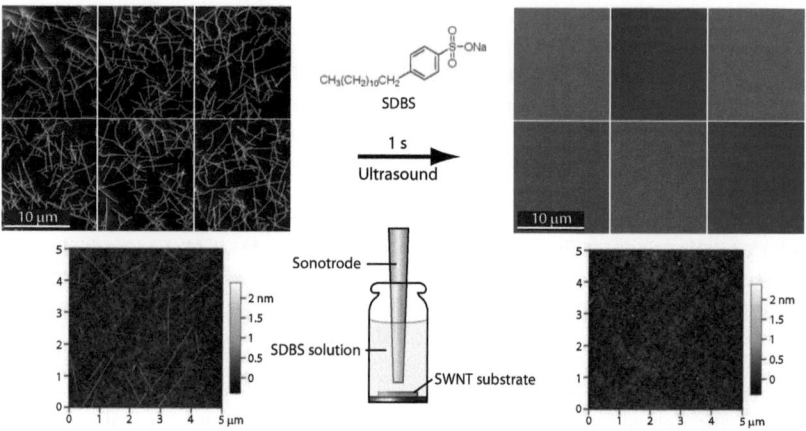

Figure 3.1: SWNT aqueous dispersion. SEM (top) and AFM (bottom) images of SWNTs grown on a SiO$_2$ substrate by a CVD process with Fe catalyst particles (left). After short, low-energy ultrasonic treatment for 1 s in a 1 wt % aqueous SDBS solution (center), the SWNTs disperse into the solution and practically none remain on the growth substrate. AFM scans reveal that the majority of catalyst particles, with a height of 2 − 3 nm, remain as white islands on the chip surface (right).

Thermal Annealing

by using an oscilloscope (LeCroy LC334A) while the chip remains grounded through the doped Si substrate. After 60 s, the generator is switched off and the chip rinsed in DIW. Finally, thermal annealing at 450 °C in nitrogen atmosphere, in order to improve the wetting interaction between the SWNTs and underlying Pd electrodes, and electric transport measurements at ambient conditions are performed on the samples to assess their electrical performance. The samples are not subjected to any further processing, such as top-side metallization, with the SWNTs simply adhering to the surface of the contact electrodes. Figure 3.2 schematically shows the employed chip design and dielectrophoretic deposition process.

SWNT Characterization

Identification of electrode bridging carbon nanotubes is performed in a scanning electron microscope at low acceleration voltages of 1 kV to minimize the risk of inducing any damage to the SWNTs during visualization [64]. On every successful deposition, electrical transport measurements are performed by applying a source-drain voltage of $V_{sd} = 60$ mV between the electrodes, while sweeping the back gate potential from $V_g = -10$ V to $V_g = +10$ V in a pulsed measurement setup. Ideal metallic SWNTs do not evidence any gate dependency, whereas semiconducting SWNTs behave like p-type field effect transistors under ambient conditions [23]. Dis-

3.3 Results

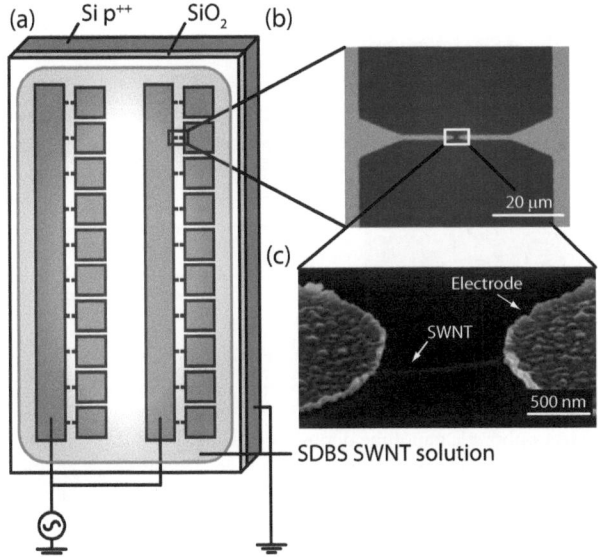

Figure 3.2: Schematic and micrographs of the DEP chip design. (a) Electrical connection scheme adopted for the dielectrophoretic deposition of individual SWNTs. (b) Optical microscopy image of an electrode pair. (c) SEM image of a successfully DEP deposited SWNT on an electrode.

crimination between metallic and semiconducting single-walled carbon nanotubes is made at a threshold on/off ratio of 3. The on/off ratio is defined as the quotient of the average current measured between gate voltages from -6 to -10 V and $+6$ to $+10$ V.

3.3 Results

Results show that the initially straight and individually distinguishable carbon nanotubes, with few twists and bends after growth (Figure 3.1), readily disperse into the aqueous SDBS stabilized solution. After being subjected to the earlier described process, the substrate images evidence essentially no presence of leftover nanotubes in scanning electron microscopy (SEM) and atomic force microscopy (AFM) images. Only very few remaining tubes can be found following the low energy ultrasonic

Ultrapure Solutions

3 Aqueous Dispersion and Dielectrophoretic Assembly of Individual SWNTs

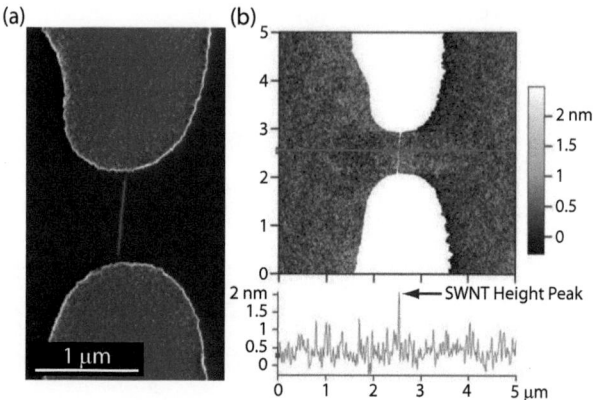

Figure 3.3: Dielectrophoretic nanotube deposition. (a) SEM and (b) AFM image with height profile, showing the successful dielectrophoretic deposition of an individual SWNT from the prepared solution. The diameter of the SWNT measured by AFM is 2 nm.

input. Iron catalyst particles, however, remain in large number on the silicon oxide surface, as seen in AFM scans. This suggests ultrapure SWNT aqueous solutions, requiring no additional purification steps.

SWNT Deposition   The directed dielectrophoretic deposition of individual SWNTs is confirmed by SEM and AFM micrographs (Figure 3.3). The length of the deposited tubes is in the same size range as prior to the dispersion, remaining on average about 2 μm long. In addition, practically no bundles are apparent in the performed deposition experiments. The solution largely contains individually dispersed SWNTs, which do not agglomerate during their surface detachment and are stabilized by the encapsulating SDBS micelle. Yields of 30% individual SWNTs bridging electrode gaps after dielectrophoresis are obtained, with a maximum yield of 55% being achieved for one individual chip. All experiments were performed over a time span of 6 months, proving the long-term stability of the prepared SWNT solutions.

Electrical Behavior   Successful dielectrophoretic deposition was achieved for 223 electrode gaps on which electrical transport measurements were performed. Metallic nanotube behavior according to the introduced criteria above was observed in 73 cases, versus 62 cases for semiconducting nanotube behavior. All bridged gaps exhibited a measurable electrical signal. However, inconclusive characteristics in 88 instances resulted in a 60% unambiguous and well-defined electrical characterization yield. Figure 3.4 exemplifies the behavior of a metallic and semiconducting carbon nanotube, along with the accompanying box plots of the average resistance for metallic tubes and

3.3 Results

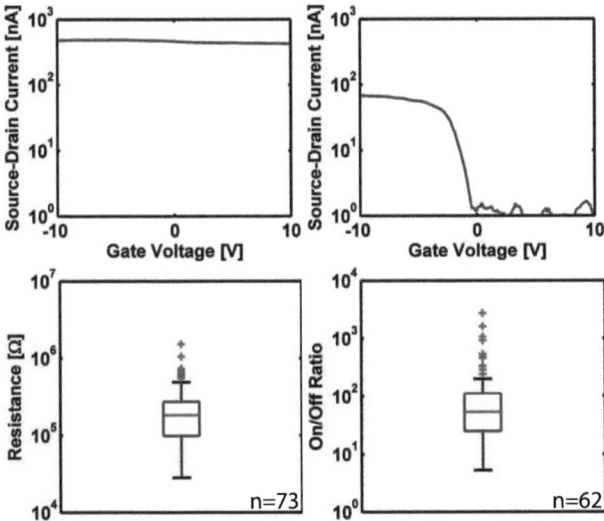

Figure 3.4: Examples of metallic and semiconducting single-walled carbon nanotube behavior along with the corresponding box plots for system resistances of metallic and on/off ratios of semiconducting SWNTs. The cutoff value between metallic and semiconducting behavior is performed at an on/off ratio of 3. The source-drain voltage is set to $V_{sd} = 60$ mV.

on/off ratios for semiconducting tubes. The central mark in the boxes is the median, the edges are the 25th and 75th percentiles, while the whiskers extend to the most extreme data points (outliers not considered). The maximum size of the whiskers is 1.5 times the box height. Outliers are plotted individually. The median of the electrical system resistances for metallic nanotubes is $R_{average} \approx 200$ kΩ, with values as low as $R_{low} \approx 20$ kΩ and as high as $R_{high} \approx 2$ MΩ being measured. On/off ratios for semiconducting nanotubes are on average in the range of 80, with values as low as 3 being reached, the cutoff value, and attaining maximum values up to 2000. All semiconducting SWNTs display p-type semiconducting characteristics. Linear current-voltage relationships at the on-state show no observable Schottky barrier at the metal-nanotube junction for the applied potentials. For electrode pairs with no bridging nanotubes, the observed resistance approaches infinity. All devices with integrated SWNTs retain their linear current-voltage characteristics at least 4 months after fabrication.

3 Aqueous Dispersion and Dielectrophoretic Assembly of Individual SWNTs

3.4 Discussion and Conclusions

Gentle SWNT Treatment

In comparison to the standard preparation process of individually dispersed SWNT solutions [53], the present method provides a much gentler ultrasonic treatment, as the applied energy input is mainly required to detach the SWNTs from the growth support surface and not to untangle large aggregates. The CVD grown nanotubes are in minimal contact after growth, primarily only bridging each other due to the low growth density. Bundles and ropes which would require untangling prior to the dispersion are very limited, as the catalyst particles are well distributed on the chip surface and each particle only gives rise to one SWNT growth site because of its constrained dimensions resulting from the encapsulating Ferritin protein [63]. The minimized low-energy ultrasonic treatment to detach the nanotubes from the growth support surface greatly reduces the risk of cavitation-induced scission. By subjecting the solution to maximally 25 s of ultrasound, compared to several minutes for the standard dispersion treatment, the number of defects induced on the SWNTs is expected to significantly decrease [65]. No evidence for cavitation induced scission is found when comparing nanotube lengths after growth and deposition. Further, impurities generally contained in raw bulk-synthesized HiPCO material, such as amorphous carbon, carbon nanoparticles, and interspersed Fe nanoparticles, are avoided by growing the nanotubes in a more controlled CVD process [66]. Together with the vast majority of metal catalyst particles remaining attached to the chip surface, due to the good metal-SiO2 adhesion, this eliminates the necessity of subsequent purification, especially oxidative purification processes which could induce defect formation and/or functional groups on the sidewalls of the SWNTs. In addition, without the need for centrifugation, heavy water (D_2O) can be replaced with DIW, since no density gradients are required.

Solution Concentration

Successful dielectrophoretic deposition implies that, after casting a droplet of the dispersed SWNT solution on the chip, single-walled carbon nanotubes are in close enough proximity of the locally generated electric field between the electrodes. From sample characterization after the growth process (Figure 3.1), it is estimated that the growth density lies in the range of 50 SWNTs per 100 μm^2. To evaluate the tube concentration in the solution, it is assumed that the diameter of the carbon nanotubes is about $1-3$ nm and their length on average is 2 μm, as can be inferred from AFM scans. The number of carbon atoms per SWNT can then be calculated to be around $(2.5-7.5) \times 10^5$. This results in a mass of $\sim (5-15) \times 10^{-18}$ g. By taking the 25 processed chips per solution into account, a SWNT concentration in the order of $3-9$ ng/mL is determined. This corresponds to about 1 nanotube/$(10 \mu m)^3$, slightly larger than the length scale of the gap distance, and is equivalent to 3×10^6 nanotubes per droplet used in the dielectrophoretic deposition process. As these solution concentrations are too low to perform UV-vis absorption spectrometry or

3.4 Discussion and Conclusions

spin-coating on a substrate, dielectrophoresis along with electrical characterization permits one to confirm that individually dispersed single-walled carbon nanotubes are present within the prepared solutions.

With deposition yields in the range of 30% for individually bridging tubes when employing optimized deposition parameters, the integration of SWNT-based systems by purely parallel processes is demonstrated. The deposited tubes are in general straight, thus minimizing local band gap opening by bending [67]. For the unsuccessful deposition cases, no nanotubes at all are observed, the SWNTs are only attached to one electrode, or multiple distinguishable individual SWNTs bridge the gaps. Smaller electrodes and gap distances are expected to increase the deposition yield due to stronger and more localized electric field gradients and may even enable a self-limiting assembly mechanism [68]. With the appropriate surfactant choice, the long-term stability of the suspensions is well guaranteed and observed. This allows the prepared solutions to be used in excess of 6 months. By rinsing off nondeposited tubes after dielectrophoresis in deionized water, before drying the chips in a stream of nitrogen gas, toxicological risks are kept to a minimum, because the nanotubes never become airborne.

DEP Yield

Long-Term Stability

The quality of single-walled carbon nanotube solutions and raw material is assessed by their electrical characterization. Two-point system resistances are on average 1 order of magnitude lower than those of similar studies employing HiPCO nanotube material [68]. Resistances in the present work reach values almost as low as resistances after an additional top-side metallization in the previously mentioned study. This is assumed to be due to the generally larger tube diameters resulting from the CVD growth process (1 − 3 nm), since SWNT resistances are diameter dependent [69], as well as due to the reduced defect density resulting from the limited processing, in particular low-energy ultrasonic input and avoidance of (oxidative) purification processes. The low resistance is further attributed to the high work function and good wetting interactions of the palladium electrodes with the nanotubes [70]. The thermal annealing step additionally enhances the wetting of the metal-nanotube interface and desorption of residual surfactant from the nanotube surface, thus further reducing the resistance in the system. The comparatively low on/off ratios for semiconducting SWNTs are ascribed to the larger diameter distribution of the CVD grown nanotubes. As the band gap of semiconducting single-walled carbon nanotubes scales inversely with tube diameter, Schottky barrier heights are reduced, thus increasing the subthreshold current in the device. Long-term electrical measurements, without any additional passivation provisions, are taken as an indication that no significant barriers are formed at the Pd-SWNT contact interfaces. The nanotubes adhere permanently on the electrodes and are not even detached by high intensity ultrasonication. The excellent contact interface is confirmed by the exhibited electrical signals for all deposited nanotubes. Electrode contacting is fur-

Process Improvements

3 Aqueous Dispersion and Dielectrophoretic Assembly of Individual SWNTs

ther improved by the long contact area between the nanotubes and the electrodes. Electrical nanotube characteristics which do not follow metallic or semiconducting SWNT behavior may be attributed to roped SWNTs, which cannot be completely excluded from the solution. Since the magnitude of the dielectrophoretic force depends on the volume of the trapped object, this causes their preferential deposition. The highly unambiguous electrical characterization yield of 60% for all individually deposited nanotubes is a strong indication, however, of the high quality SWNT raw material, solution, and contact interface.

Conclusions   The herein introduced method provides an important approach to limit cavitation-induced scission and defect inclusion in preparing individually dispersed SWNT solutions. By considerably reducing the amount of ultrasonic energy input, the practically bundle-free surface-synthesized CVD tubes remain longer and appear to be more rigid than bulk-synthesized HiPCO tubes. Further, electrical characterization at low resistances demonstrates the high quality of the SWNT solutions, raw material, and contact interface. These long-term stable ultrapure solutions, with long and largely defect-free SWNTs, may therefore be of significant interest to the research community and enable further developments in directed carbon nanotube device assembly techniques, combined with selective SWNT growth or dispersion, for device and sensor integration.

4 Selective Parallel Integration of Individual Metallic Single-Walled Carbon Nanotubes from Heterogeneous Solutions

Parts of this chapter are published in:

B. R. Burg, J. Schneider, V. Bianco, N. C. Schirmer & D. Poulikakos. Selective parallel integration of individual metallic single-walled carbon nanotubes from heterogeneous solutions. *Langmuir* **26**, 10419-10424 (2010).

http://pubs.acs.org/articlesonrequest/AOR-UhwCaea4vugBcukV5IHA

Abstract

The dielectrophoretic separation of individual metallic single-walled carbon nanotubes (SWNTs) from heterogeneous solutions and their simultaneous deposition between electrodes is achieved and confirmed by direct electric transport measurements. Out-of-solution guided parallel assembly of individual SWNTs was investigated for electric field frequencies between 1 and 200 MHz. At 200 MHz, 19 of the 22 deposited SWNTs (86%) displayed metallic behavior, whereas at lower frequencies the expected random growth distribution of 1/3 metallic SWNTs prevailed. A threshold separation frequency of 188 MHz is extracted from a surface-conductivity model, and a conductivity weighting factor is introduced to elucidate the separation frequency dependence. Low-frequency experiments and numerical simulations show that long-range nanotube transport is governed by hydrodynamic effects whereas local trapping is dominated by dielectrophoretic forces. The electrokinetic framework of dielectrophoresis in low-concentration solutions is thus provided and allows a deeper understanding of the underlying mechanisms in dielectrophoretic deposition processes for long and large-diameter SWNT-based low-resistance device integration.

4 Selective Parallel Integration of Individual mSWNTs from Heterogeneous Solutions

4.1 Introduction

SWNT Separation

The promise of single-walled carbon nanotubes (SWNTs) as inherent components in nanoscale electronic devices, sensors, and applications has inspired a large range of research on these quasi-1D nanostructures [71]. Like other nanomaterials, the electric and optical properties of SWNTs depend on their size and atomic structure, known as chirality [18]. The limitations in integrating SWNTs according to their electric properties [56] while entirely relying on parallel fabrication techniques have, however, constrained their emergence in commercial applications to date. Metallic SWNTs are thought to take the role of leads in nanoscale circuits, and for the realization of nanotube-based electronics, it is essential to manipulate metallic and semiconducting SWNTs separately. Dielectrophoresis (DEP) [11], allowing the site-selective deposition [32, 60] as well as the conductivity-dependent separation of SWNTs [35], provides this prospect in a single-step process [30, 51, 52]. The research reported herein demonstrates the highly selective parallel assembly of individual metallic SWNTs onto prefabricated electrodes from heterogeneous solutions by dielectrophoresis and introduces the underlying electrokinetic framework of the deposition and separation process.

4.2 Experimental Section

SWNT Dispersion

SWNT solutions were prepared by removing surface-synthesized carbon nanotubes from a growth-support substrate by a short, low-energy ultrasonic pulse to produce ultrapure, long-term-stable 1 wt% sodium dodecylbenzenesulfonate (SDBS) surfactant-stabilized aqueous solutions, according to a previously published account [34]. The surfactant concentration is 1 order of magnitude above the critical micelle concentration (cmc) to ensure successful SWNT dispersion [72]. The SWNTs were grown by a well-documented iron (Fe)-catalyst-based chemical vapor deposition (CVD) process to yield carbon nanotubes with a comparatively large mean diameter of 1.9 ± 0.8 nm, containing all chiralities from this diameter range and exhibiting a low defect density, as evidenced by Raman spectroscopy [63, 73]. In comparison to the standard preparation method of individually dispersed SWNT solutions from bulk-synthesis high-pressure CO (HiPCO) reactor processes [53], the process induces much gentler ultrasonic treatment. This is expected to decrease the number of defects on the SWNTs [65].

SWNT DEP

Dielectrophoretic deposition of individual SWNTs was performed on chips containing 20 individually accessible electrode gaps in the range of 1 to 2 μm (typical parameters: V_p = 2 V, t = 60 s) [34]. An insulating dry thermal oxide layer allowed capacitive coupling between the p-type doped silicon substrate and palladium (Pd)

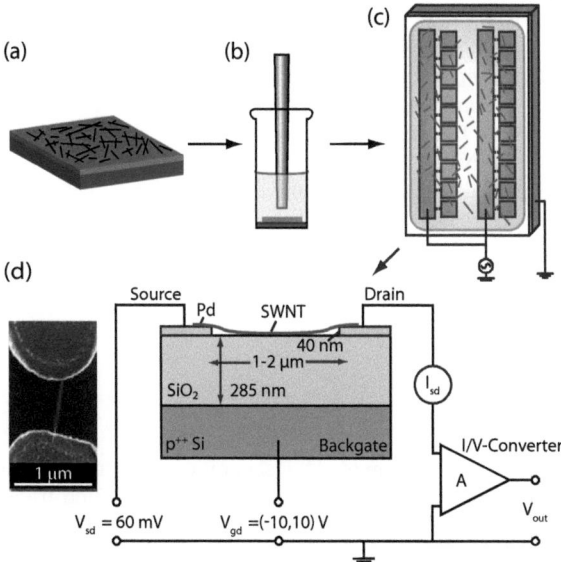

Figure 4.1: Experimental methods. (a) Surface-synthesized SWNTs grown in a CVD oven. (b) Aqueous dispersion of SWNTs in surfactant-stabilized solutions. (c) Dielectrophoretic deposition of individual SWNTs by capacitive coupling. (d) Topographical and electrical characterization of deposited carbon nanotubes.

electrodes for large-scale assembly [31] as well as electric field-effect-transistor characterization [74, 75]. After dielectrophoresis, the chips were rinsed in deionized water (DIW) and thermally annealed at 450 °C in a nitrogen atmosphere to remove any residual surfactant wrapping the carbon nanotubes [76] and to improve the wetting interactions between the SWNTs and the underlying Pd electrodes [70]. Electrical characterization of individual SWNTs, identified by scanning electron microscopy (SEM) and atomic force microscopy (AFM), was carried out by linear gate sweeps ($V_g = \pm 10$ V) at a constant source drain voltage ($V_{sd} = 60$ mV) [77]. For an on/off ratio larger than 3, the SWNTs were considered to be semiconducting. Detailed electrical transport data of dielectrophoretically integrated individual SWNTs from identical solutions are reported in a prior study [34]. Among others, the electrical system resistances of metallic SWNTs were measured to average $R_{av} \approx 200$ kΩ. Linear current-voltage relationships of semiconducting SWNTs in the on state showed no observable Schottky barriers at the metal-nanotube junction for the applied po-

Electrical Characterization

4 Selective Parallel Integration of Individual mSWNTs from Heterogeneous Solutions

Deposition Study
tentials. Low-resistance devices and a high electrical characterization yield verified the high quality of the SWNT raw material, individually dispersed solutions, and integrated contact interface.

The frequency-dependent dielectrophoretic separation of SWNTs was investigated at varying deposition frequencies ranging from 1 to 200 MHz. For random growth, a 1/3 metallic (mSWNT), 2/3 semiconducting (sSWNT) single-walled carbon nanotube ratio is expected [78]. To determine the prevailing distribution ratio, multiple SWNTs must be electrically probed and statistical distributions must be studied. For an ideal distribution, the Bernoulli distribution applies and a one-sided binomial (Bin) test can be performed. The null hypothesis of a $p = 2/3$ sSWNTs ratio is violated within a confidence level of 99% when the sSWNT ratio drops below 40% for $n = 20$ randomly probed carbon nanotubes.

4.3 Results

mSWNT Integration
Figure 4.2 shows the results from the DEP separation experiments. The ratio of semiconducting carbon nanotubes is plotted against the deposition frequency, with the number of electrically characterized SWNTs shown at each data point. For frequencies between 10 and 80 MHz, the expected random growth ratio of 2/3 sSWNTs was confirmed, ranging from 52 to 68%. At 100 MHz, the obtained sSWNT ratio was 23% and only 3 of the 22 deposited SWNTs at 200 MHz showed semiconducting behavior (14%). This accounts for a statistically relevant and significant enrichment of mSWNTs at higher deposition frequencies, as is illustrated by the location of the data points in the critical region of the one-sided binomial test Bin(20,2/3). Additionally, an enrichment of mSWNTs at 1 MHz is observed. The feasibility to manipulate and integrate individual metallic SWNTs selectively from heterogeneous solutions is thus demonstrated.

4.4 Discussion and Conclusions

Electrokinetics
The electrokinetic understanding of the system provides the basis for explaining these observations.

DEP Force
The time-averaged dielectrophoretic force on a rod-shaped particle with its major axis parallel to the electric field lines equals [79]

$$\langle \mathbf{F_{DEP}} \rangle = \frac{\pi abc}{3} \epsilon_m \mathrm{Re} \left\{ \frac{\epsilon_p^* - \epsilon_m^*}{\epsilon_m^*} \right\} \nabla \left| \mathbf{E} \right|^2, \tag{4.1}$$

with a, b, and c being the half-lengths of the major ellipsoid axes, ϵ_p^* and ϵ_m^* being the complex permittivity of the particle and suspension medium, respectively, and $\mathbf{E}$

4.4 Discussion and Conclusions

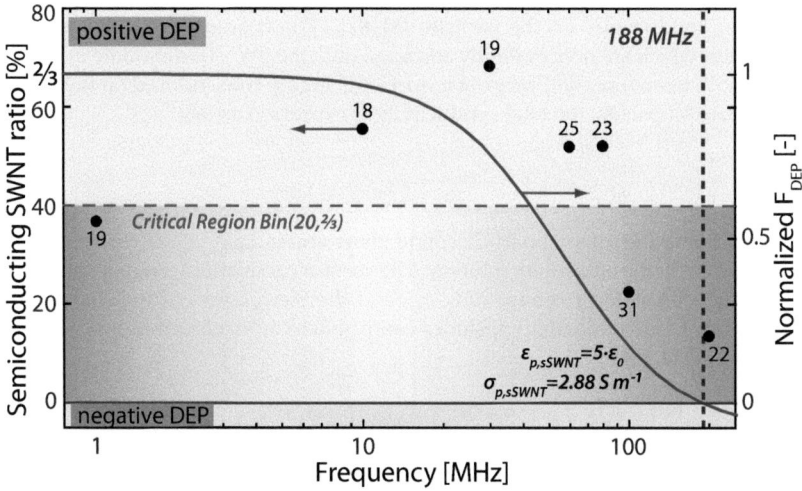

Figure 4.2: Dielectrophoretic deposition results for frequencies between 1 and 200 MHz. The ratio of semiconducting carbon nanotubes is plotted against the deposition frequency, with the number of electrically characterized SWNTs shown at each data point. On the right axis, the normalized DEP force for sSWNTs with a surface conductivity of $\sigma = 2.88$ S m^{-1} and a permittivity of $\epsilon = 5\epsilon_0$ is displayed. A crossover from positive to negative dielectrophoretic force orientation is observed at 188 MHz, preceded by a continuous decrease in magnitude. The critical region of the one-sided binomial test Bin(20,2/3) with a confidence level of 99% shows where a statistically significant enrichment of metallic carbon nanotubes is found in the dielectrophoretic deposition of SWNTs from heterogeneous solutions. DEP explains the highly selective integration of mSWNTs at high frequencies (100 − 200 MHz). The enrichment of mSWNTs at 1 MHz is explained by the hydrodynamic behavior of the system.

being the electric field. For a constant particle size in a given solution, the magnitude of the dielectrophoretic force at a specific location solely depends on the complex permittivity of the particle $\epsilon_p^* = \epsilon_p - i\frac{\sigma_p}{\omega}$, where ϵ is the permittivity, σ is the conductivity, and ω is the angular frequency of the electric field [79].

The permittivity of SWNTs is inversely proportional to the square of the energy band gap [80]. Consequently, the permittivity of mSWNTs is exceedingly large, whereas for sSWNTs a value of $5\epsilon_0$ or less is generally adopted [32]. The inherent conductivity of mSWNTs causes a considerable particle conductivity, and sSWNTs are expected to experience zero intrinsic conductivity because of their band gap.

sSWNT Permittivity

4 Selective Parallel Integration of Individual mSWNTs from Heterogeneous Solutions

Surface Conductivity Model
The dispersion of SWNTs in surfactant-stabilized solutions, however, gives rise to a surface conductivity on the particles [81, 82]. The headgroups of the surfactant molecules, which are noncovalently adsorbed onto the SWNTs, dissociate in aqueous solution. A temporary and removable surface charge is thus induced on the SWNTs. For prolate ellipsoids, the total conductivity is expressed as [83]

$$\sigma_p = \sigma_{int} + \frac{2\sigma_s}{b} = \sigma_{int} + \frac{2(\sigma_{s,d} + \sigma_{s,s})}{b}, \quad (4.2)$$

with σ_{int} being the internal particle conductivity and σ_s being the developed surface conductivity in the electrolyte solution. The surface conductivity is composed of two terms, the diffuse layer conductivity $\sigma_{s,d}$ and the Stern layer conductivity $\sigma_{s,s}$ [84]. The diffuse layer conductivity includes contributions from electroosmosis and ionic conduction [14, 43]:

$$\sigma_{s,d} = \frac{4qN_A c_0 z^2 \mu}{\kappa}\left(1 + \frac{2\epsilon_m k_B T}{z^2 \mu \eta q}\right)\left(\cosh\left(\frac{zq\zeta}{2k_B T}\right) - 1\right), \quad (4.3)$$

with all of the variables according to Table 4.1. The Stern layer conductivity is assumed to be proportional to the diffuse layer conductivity [85, 86]:

$$\sigma_{s,s} = 0.56 \cdot \sigma_{s,d}. \quad (4.4)$$

sSWNT Conductivity
Calculating the surface conductivity of sSWNTs, which corresponds to the total conductivity where the internal particle conductivity σ_{int} equals zero, yields $\sigma_{p,sSWNT} = 2.88$ S m^{-1}. By assuming a permittivity of $\epsilon_{p,sSWNT} = 5\epsilon_0$ for sSWNTs, we can estimate the magnitude of the dielectrophoretic force acting on the semiconducting carbon nanotubes. The right axis of Figure 4.2 displays the normalized values of the DEP force acting on semiconducting SWNTs.

DEP Behavior
A crossover from positive to negative dielectrophoretic force orientation is observed, preceded by a continuous decrease in magnitude. The crossover frequency is evaluated to occur at 188 MHz, above which no dielectrophoretic deposition of sSWNTs is expected. The experimental results overlaid in the graph clearly confirm the trend of decreasing semiconducting carbon nanotube ratio at a deposition frequency of 100 MHz and even more significantly at 200 MHz. This is taken as strong evidence of the decreasing dielectrophoretic force magnitude acting on the sSWNTs in the heterogeneous solution, and it is concluded that dielectrophoresis is the underlying mechanism for the separation of SWNTs at high frequencies.

Clausius-Mossotti Factor
Figure 4.3 elucidates the origin and background of the dielectrophoretic crossover frequency. Particles in a solution move toward regions of highest electric field strength if the polarizability of the particles is greater than that of the suspend-

4.4 Discussion and Conclusions

Variable	Description	Value	Reference
c_0	Electrolyte concentration	28.696 mol m^{-3}	
k_B	Boltzmann constant	$1.38 \cdot 10^{-23}$ J K^{-1}	[45]
$l(=2c)$	SWNT length	$2.5 \cdot 10^{-6}$ m	
N_A	Avogadro's constant	$6.022 \cdot 10^{23}$ mol^{-1}	[45]
q	Elementary charge	$1.60 \cdot 10^{-19}$ C	[45]
$2r(=2a=2b)$	Diameter of micellized SWNTs	$3 \cdot 10^{-9}$ m	
T	Temperature	298 K	
z	Valence of Na$^+$	1	[45]
ϵ_0	Vacuum permittivity	$8.854 \cdot 10^{-12}$ F m^{-1}	[45]
ϵ_m	Water permittivity	$80\epsilon_0$	[45]
ζ	Zeta (ζ)-potential	-60.4 mV	[87]
η	Water viscosity	$0.890 \cdot 10^{-3}$ Pa s	[45]
$\kappa = \frac{2z^2 q^2 N_A c_0}{\epsilon_m k_B T}$	Inverse Debye length	calculated	[43]
μ	Mobility of Na$^+$	$5.19 \cdot 10^{-8}$ m^2 (V s)$^{-1}$	[14]
σ_m	Solution conductivity (1 wt % SDBS)	0.250 S m^{-1}	measured

Table 4.1: Variables and corresponding values used for determining the surface conductivity of sSWNTs and the electrothermal flow in the suspension medium.

ing medium; otherwise, they are repelled. The polarizability ratio is frequency-dependent and described by the Clausius-Mossotti factor [79]

$$f_{CM} = \left\{ \frac{\epsilon_p^* - \epsilon_m^*}{\epsilon_m^*} \right\}. \tag{4.5}$$

In the low-frequency limit, the real part of the Clausius-Mossotti relation depends solely on the conductivities of the particle and the suspending medium. Conversely, in the high-frequency limit the polarization is dominated by the permittivities of the particle and the suspending medium. A change in the polarizability of the particles with respect to the suspending medium at a particular crossover frequency causes a change in the sign of the Clausius-Mossotti factor and consequently an inversion of the dielectrophoretic force direction.

The variation of the crossover frequency of sSWNTs as a function of solution conductivity for a fixed particle surface conductivity ($\sigma_{p,sSWNT} = 2.88$ S m^{-1}) is depicted in Figure 4.3(a). For positive, attractive dielectrophoresis to occur in the low-frequency region, the particle conductivity must be greater than the solution conductivity. Once the solution conductivity exceeds the particle conductivity, the dielectrophoretic force becomes negative and thus repelling. In the high-frequency region, where the dielectric constant in the aqueous suspension solution always

4 Selective Parallel Integration of Individual mSWNTs from Heterogeneous Solutions

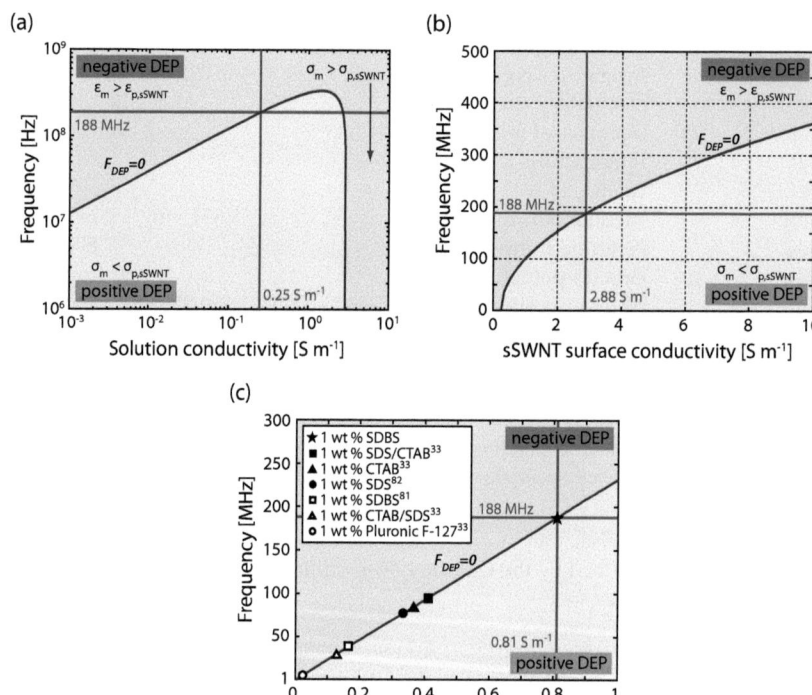

Figure 4.3: Background on the dielectrophoretic crossover frequency for sSWNTs. (a) Solution conductivity dependency on the dielectrophoretic crossover frequency for sSWNTs with a surface conductivity of $\sigma_{p,sSWNT} = 2.88$ S m^{-1}. (b) sSWNT surface conductivity dependency on the dielectrophoretic crossover frequency in a solution with a conductivity of $\sigma_m = 0.250$ S m^{-1}. (c) Linear influence of the conductivity weighting factor $\gamma = \sqrt{\sigma_p \sigma_m - \sigma_m^2}$ on the sSWNT crossover frequency. Different SWNT solutions from the literature are added to the representation, illustrating the benefit of the introduced parameter.

exceeds the dielectric constant of the sSWNTs, the dielectrophoretic force remains negative. For the present experimental conditions ($\sigma_m = 0.250$ S m^{-1}) the identified crossover point from positive to negative dielectrophoresis at 188 MHz, where the polarizability of the sSWNTs and the solution equal each other, is marked by crosslines.

The particle surface conductivity, however, also play a primary role in the location of the crossover frequency. Figure 4.3(b) shows this dependency for a fixed solution conductivity ($\sigma_m = 0.250$ S m^{-1}). The current conditions ($\sigma_{p,sSWNT} = 2.88$ S m^{-1}) are again marked by crosslines and the interlinked decrease of particle surface conductivity and dielectrophoretic crossover frequency behavior becomes apparent. Without a negative surface charge on semiconducting SWNTs no dielectrophoretic deposition is expected to occur, as the dielectrophoretic force always remains negative. Particle Conductivity

By introducing the conductivity weighting factor $\gamma = \sqrt{\sigma_p \sigma_m - \sigma_m^2}$, which originates from finding the roots of the Clausius-Mossotti factor, the combined effects of particle surface conductivity and solution conductivity on the crossover frequency can be investigated. Figure 4.3(c) shows that the crossover frequency depends linearly on the conductivity weighting factor γ. This factor must be reduced to scale down the crossover frequency proportionally and may be tuned according to specific requirements. The largest influence resides in the particle surface conductivity, which is dependent on the ζ-potential and can be influenced by adjusting the pH-value of the solution [87], by adding a cationic surfactant [82,88] or by using nonionic surfactants [33]. Related studies specifically investigate this aspect [82,89]. A certain dependency of the solution conductivity on the particle surface conductivity, however, always remains. Different SWNT solutions from the literature with varying solution and surface conductivity values are displayed in Figure 4.3(c) and show the benefit of the introduced factor γ. Conductivity Weighting Factor

To complete the picture, the dielectrophoretic force of mSWNTs always remains positive and is at considerably higher values than for sSWNTs. This is because mSWNTs are much more polarizable than the surrounding medium under all circumstances. Conductivity and permittivity values of $\sigma_{p,mSWNT} = 10^3$ S m^{-1} and $\epsilon_{p,mSWNT} = 10^3 \epsilon_0$ are adopted for mSWNTs in this study. mSWNTs

Previous reports on the dielectrophoretic separation of SWNTs were all performed on thin films. Different surfactants led to a shift in the dielectrophoretic crossover frequency from below 10 MHz in 1 wt % SDS (sodium dodecyl sulfate) [35] to 30 – 40 MHz in 1 wt % SDBS [81] and 50 – 70 MHz in 1 wt % NaCh (sodium cholate) aqueous solutions [90] on samples analyzed by Raman spectroscopy. The dielectrophoretic separation of SWNTs has further been reported on thin films below 10 MHz for 1 wt % CTAB/SDS (cetylrimethylammonium bromide) solutions examined by Raman spectroscopy and electrical characterization [33, 82, 88]. The Literature

4 Selective Parallel Integration of Individual mSWNTs from Heterogeneous Solutions

comparatively low separation frequencies may be explained by the small sSWNT residual in the deposited film that cannot always be resolved by the Raman and/or electrical signal and consequently distorts the extracted crossover values. No separation from other solutions has been shown to date, because most do not provide the required complete dispersion of SWNTs [33]. Continuous extraction of mSWNTs in a microfluidic channel in 1 wt % CTAB has also been published at an applied frequency of 10 MHz [91]. The setup, however, relied on the much larger dielectrophoretic attraction force of mSWNTs with respect to that of sSWNTs and the involved semiconducting carbon nanotubes still experienced a positive DEP force under the conditions that were used, yet this force was much lower in magnitude.

Hydrodynamics

To explain the enrichment of mSWNTs at 1 MHz, the hydrodynamic behavior of the system must be studied. The applied electric field causes Joule heating in the electrolyte solution. Local heating of the fluid gives rise to permittivity and conductivity gradients, which in turn generate a body force on the fluid. This electrothermal force, written in terms of the temperature gradient, is equal to [41]

Electrothermal Force

$$\mathbf{f_e} = \frac{1}{2}\text{Re}\left[\frac{\sigma_m \epsilon_m (\alpha - \beta)}{\sigma_m + i\omega\epsilon_m}(\nabla T \cdot \mathbf{E})\mathbf{E}^* - \frac{1}{2}\epsilon_m \alpha |\mathbf{E}|^2 \nabla T\right] \tag{4.6}$$

with $\alpha = (1/\epsilon_m)(\partial \epsilon_m/\partial T) \approx -0.4\%~\text{K}^{-1}$ [45] and $\beta = (1/\sigma_m)(\partial \sigma_m/\partial T) \approx 2\%~\text{K}^{-1}$ [45]. At low-frequencies, the first term in the equation - the Coulomb force - dominates and at high-frequencies, the second term - the dielectric force - takes over. The two forces act in different directions over a certain range of frequencies, resulting in a changing flow pattern.

Numerical Simulation

By solving the coupled quasi-static Maxwell equations, heat equation, and Navier-Stokes equation with the added body force for an applied voltage of $V_p = 2$ V and a 1 µm electrode gap in a commercial software tool (COMSOL MULTIPHYSICS v3.5a), the impact of the electrothermal flow (ETF) is investigated [37,38]. The model allows the dielectrophoretic force to be superimposed and its effect on massless rod-shaped particles with a corresponding friction factor of $f = 3\pi\eta l/\ln(l/r)$ to be determined [14].

Electrokinetic Behavior

Figure 4.4 shows the results of the numerical simulations for applied frequencies of 1, 60, and 200 MHz. The surface and arrow plots display the magnitude and direction of electrothermal flow in the system. At the lower frequencies, two vortices are observed, and a change in flow properties at higher frequencies occurs, where the dielectric component of the electrothermal force takes over. The magnitude of the ETF initially increases before decreasing again toward the change in flow pattern. This is illustrated by the varying arrow thickness in the schematic representations below the simulation results. The superimposed dielectrophoretic force on the SWNTs is displayed in the simulation results in the region where the DEP-induced particle velocity exceeds the maximum fluid velocity in the vicinity of the

4.4 Discussion and Conclusions

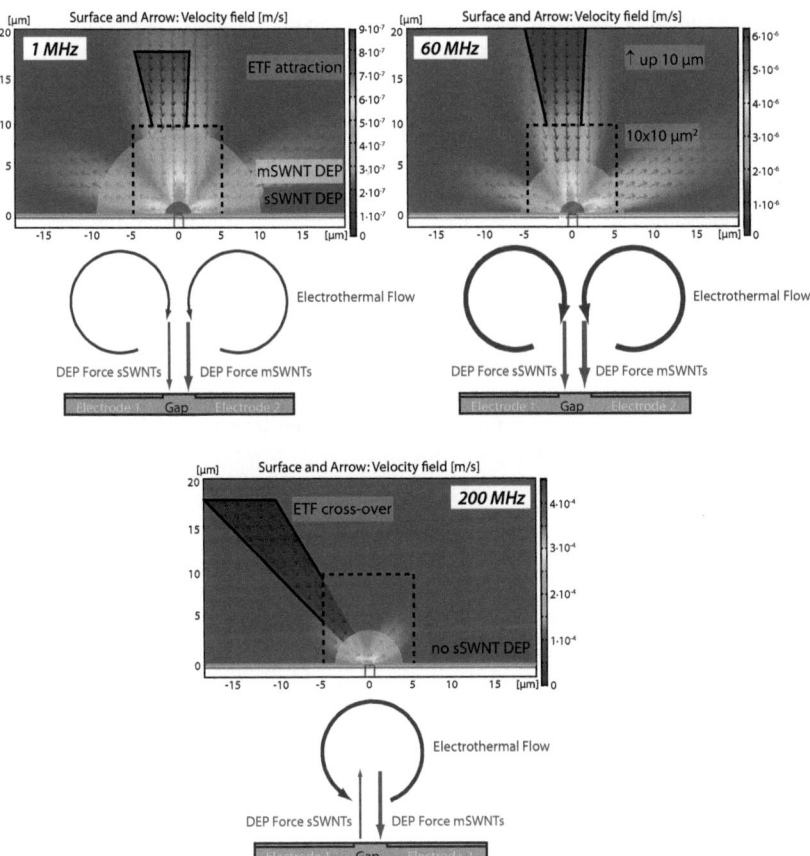

Figure 4.4: Numerical simulations of the electrothermal flow and superimposed dielectrophoretic force induced on SWNTs at 1, 60, and 200 MHz. The schematics show the comparative magnitudes of the respective forces and profile of the ETF. The dielectrophoretic attraction of mSWNTs extends much farther into the solution than for sSWNTs, although it decreases over a range of increasing frequency because of the competing effects of the ETF. The region of $10 \times 10 \ \mu m^2$, which is expected to contain on average one SWNT, and the volume transported toward the electrodes by the ETF for 60 s are also marked. This is essential to analyzing the mixing efficiency in the system, which is required to enable equal probabilities of mSWNT and sSWNT deposition below the dielectrophoretic crossover frequency. Long-range nanotube transport is governed by hydrodynamic effects, and local trapping is dominated by dielectrophoretic forces.

51

4 *Selective Parallel Integration of Individual mSWNTs from Heterogeneous Solutions*

electrode gap. It is apparent that the DEP force is very location-dependent and that the respective DEP magnitudes for mSWNTs and sSWNTs differ considerably, as is schematically represented in the bottom half of the Figure. Additional markings are made for the $10 \times 10 ~\mu m^2$ region where only one SWNT is expected to be found in the dilute dispersion [34] and for the fluid region attracted toward the electrode gap within the applied 60 s of dielectrophoresis. Qualitatively, long-range nanotube transport is governed by hydrodynamic effects, and local trapping is dominated by dielectrophoretic forces, an observation also confirmed by related studies [38].

Mixing Efficiency
To enable an equal probability of mSWNT and sSWNT deposition according to their random growth distribution, mixing of the solution above the electrodes must be guaranteed. At 60 MHz, the mixing efficiency of the electrothermal flow is high, as witnessed by the large fluid volume attracted to the electrodes during dielectrophoretic deposition. Furthermore, the higher fluid velocity causes the dielectrophoretic force, despite its strength, to restrict its influence close to the electrodes. This provides all SWNTs an equal opportunity to be transported toward the electrodes before being ultimately dielectrophoretically attracted. The fact that approximately two-thirds of the dielectrophoretically integrated randomly grown SWNTs show semiconducting behavior in this frequency range additionally confirms that the SWNTs are individually dispersed in the solution.

1 MHz
At 1 MHz, however, mixing efficiency is greatly reduced. Much less fluid volume is transported toward the electrodes, and the dielectrophoretic force effect on metallic nanotubes extends much farther into the surrounding solution. If both a metallic and a semiconducting SWNT are present in the determining volume for dielectrophoretic attraction, then an mSWNT has a much greater chance of being attracted, even though originally it might be farther away from the electrodes. The dielectrophoretic force range of metallic carbon nanotubes over a much larger area is therefore responsible for their preferential deposition and explains their relative enrichment at 1 MHz. Below 1 MHz, no deposition of SWNTs was achieved because of the practically absent capacitive coupling of the electrodes to the chip substrate [92] and the emergence of ac electroosmosis [14], an observation consistent with previous reports [93].

200 MHz
The electrothermal flow may also be responsible for the remaining semiconducting SWNTs deposited at 200 MHz, above the determined dielectrophoretic crossover frequency. If the drag experienced by SWNTs succeeds in overcoming the repulsive dielectrophoretic force, then the carbon nanotubes could contact the electrodes from where they are not able to be removed anymore [34].

Conclusions
In conclusion, the selective manipulation and direct integration of individual metallic SWNTs from heterogeneous solutions have been demonstrated and the physics behind the responsible mechanisms over the entire frequency range have been examined and explained. Electrokinetic effects, more specifically, the dielectrophoretic

4.4 Discussion and Conclusions

background and electrothermal flow, are essential in elucidating the enrichment of metallic single-walled carbon nanotube samples deposited by dielectrophoresis. The observations made by electric transport measurements on individual SWNTs match very well with a theoretical model used to determine their surfactant-induced surface conductivity. The selective integration of large-diameter, long SWNTs is a vital aspect in low-resistance SWNT sensor integration, where ohmic contacts with good contact interfaces are essential. To date, density gradient ultracentrifugation, which is used in the preparation of monodisperse SWNT solutions, cannot provide such samples [94].

5 Piezoresistive Pressure Sensors with Parallel Integration of Individual Single-Walled Carbon Nanotubes

Parts of this chapter are in press as:

B. R. Burg, T. Helbling, C. Hierold & D. Poulikakos. Piezoresistive pressure sensors with parallel integration of individual single-walled carbon nanotubes. *Journal of Applied Physics* in press (2011).

Abstract

Selective dielectrophoretic integration of individual small band gap semiconducting single-walled carbon nanotubes (SGS-SWNTs) is employed herein to enable the purely parallel fabrication of SWNT-based piezoresistive pressure sensors. Directed assembly allows precise carbon nanotubes positioning on the designated silicon oxide (SiO_2) membrane edges, ultimately the positions of highest strain. Strain components other than from the principal axis are minimized through good alignment. The SWNTs are encapsulated by a protective alumina (Al_2O_3) coating and can be modulated by a top gate. The pressure sensors have a membrane diameter of 120 μm and thickness of 190 nm. Highest sensitivity of the long-term stable devices is achieved in the off-state of SGS carbon nanotube transistors (CNFETs), reaching values as high as $S_0 \sim 0.25$ $\Delta R/R$/bar, at a resolution better than 50 mbar, and a power consumption of less than 40 nW. Low contact resistances and high transmission are essential for good sensor resolution. The scale-up of the introduced robust and reliable fabrication process is straight forward and may provide interesting avenues toward SWNT sensor commercialization.

5 Piezoresistive Pressure Sensors with Parallel Integration of Individual SWNTs

5.1 Introduction

Pressure Sensors
Piezoresistivity based pressure sensors consist of two essential components: a membrane and a transducer element, which responds to applied strain through a resistance change [95]. Traditionally, both elements are made up of silicon. A silicon-diaphragm pressure sensor consists of a thin single-crystal silicon membrane as elastic material and a piezoresistive gauge resistor created by diffusive impurities [96,97]. State-of-the-art membrane dimensions are 250×250 μm^2 in size with gauge factors for semiconducting strain gauge elements in the range of $40 - 200$ [95]. In an effort to significantly reduce the size of microelectromechanical systems (MEMS) toward realizing nanoelectromechanical systems (NEMS), single-walled carbon nanotubes (SWNTs) are anticipated to take over as transducer element for different sensing applications, exploiting their chemical, optical, resonant or electromechanical properties [21,98]. In particular, straining SWNTs modifies their electronic band gap [99]. An electrical response change may therefore be used for different sensor concepts ranging from strain, torsion, force, and pressure, amongst others. Electrical confirmation of the strain induced band gap change in SWNTs has been demonstrated in different studies [77,99–106]. These reports show high promise for strain sensing applications because of the enhanced sensitivity (gauge factors up to 500), ultra-small size, and significantly reduced drive current requirements of carbon nanotubes.

SGS-SWNTs
By investigating the electromechanical properties of SWNTs adhering to thin film membranes under small strain ($\epsilon < 0.2\%$), the potential of SWNTs as functional transducer elements in classically designed piezoresistivity based pressure sensors is proven [77,102,103,105,106]. Recent studies verify that the symmetric modulation of the electronic band gap is the dominant reason for the piezoresistive behavior of SWNTs, rendering localized defects along the carbon nanotube negligable [105]. Based on the model of thermally activated transport, a strain induced band gap opening is expected to modulate the output current most in the off-state of the carbon nanotube field effect transistor (CNFET). In the off-state, the Fermi level is at the center of the band gap, which maximizes the SWNT gauge factor [99]. Small band gap semiconducting SWNTs (SGS-SWNTs), exhibiting band gaps in the order of $k_B T = 26$ meV, [107] show to be superior to semiconducting SWNTs (s-SWNTs) with respect to the absolute strain induced current change and signal-to-noise ratio (SNR) [105,106]. The reasons are that the off-state currents remain high due to the small band gap in SGS-SWNTs and that the amplitude of the low frequency noise corrupting the signal remains low. The minuscule dimensions of SWNTs additionally allow the downscaling of the membrane to a demonstrated diameter of 40 μm, or below, underlining their potential use in the next generation of pressure sensors [105].

Piezoresistors are preferably placed at the location of maximum stress/strain, gen-

5.1 Introduction

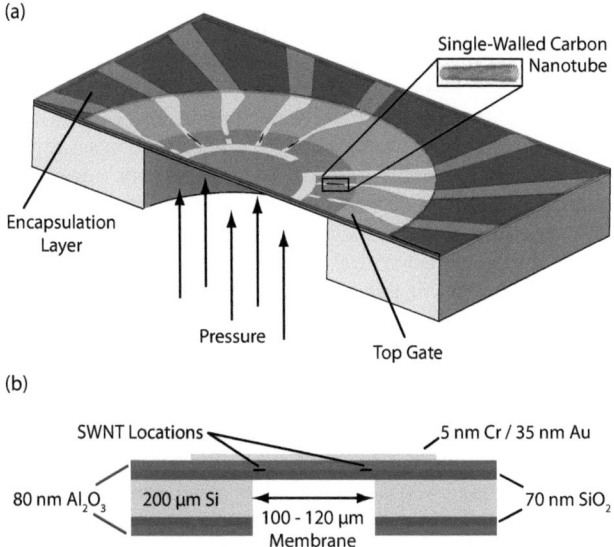

Figure 5.1: SWNT based pressure sensor schematic. (a) The individual small band gap carbon nanotubes are aligned in the radial direction on the edges of the circular membrane, the regions of maximum strain. The circular membrane is covered by a top gate and the SWNTs are protected by an encapsulating alumina layer. The radially arranged palladium electrodes are structured on silicon oxide. Purely parallel integration techniques are used in the sensor assembly. (b) Cross-section of the chip and membrane layer architecture. The SWNTs are encapsulated between a 70 nm SiO_2 and 80 nm Al_2O_3 layer. They can be modulated by a 40 nm Cr/Au top gate. The dielectric oxide layers also act as hard masks for the tri-layer 190 nm thick $100 - 120$ μm diameter membrane release.

erally the edge of a circular or rectangular membrane [108, 109]. Often also, well-matched Wheatstone bridges are configured according to precisely defined and designed positions to cancel out common-mode cross-sensitivity effects, such as temperature dependency [110]. Therefore, for commercially viable SWNT piezoresistivity based pressure sensors, one of the main challenges is precise carbon nanotube positioning. Directed assembly by dielectrophoresis (DEP) circumvents this difficulty by selectively depositing micro- and nanoscale objects from nonuniform fields onto prefabricated electrodes [11, 14, 60]. By capacitively coupling one of the electrodes in the system to a conductive substrate separated by an insulating oxide layer, minimal external contacting enables parallel large-scale assembly, while direct

Transducer Positioning

5 Piezoresistive Pressure Sensors with Parallel Integration of Individual SWNTs

current throughput through the nanostructures is avoided during the assembly process [31,32,92]. For dielectrophoretic assembly, the SWNTs have to be individually dispersed in solution, preferably employing a process which limits defect inclusion and cavitation-induced scission from extended ultrasonication [34,53].

Selective SGS-SWNT Integration

Because SGS-SWNTs exhibit the highest performance of carbon nanotube strain gauge elements, their selective integration by dielectrophoresis must be feasible. This feat is achieved by the highly selective dielectrophoretic integration of metallic SWNTs at high deposition frequencies [35,48]. According to refined carbon nanotube models, zig-zag and chiral metallic SWNTs exhibit a small band gap from curvature effects due to the cylindrical structure of the hexagonal carbon backbone structure [107]. The model of thermally activated transport, describing the SGS-SWNT band gap opening imposed by strain, specifies that the gauge factor is not only dependent on the SWNT chirality, but also on the contact properties and device transmission [98,99]. The contact resistance of SWNT devices depends on the SWNT diameter and Schottky barriers. Contact barrier heights are dependent on the band gap, reducing resistances accordingly for large-diameter SWNTs with smaller band gaps [69,111,112]. Further, the height of Schottky barriers is reduced by the appropriate choice of metal. Palladium (Pd) provides low-resistance contacts, due to its high work function and good wetting interaction with the carbon nanotubes [70,112,113]. Channel lengths longer than the depletion width of the device are additionally necessary to avoid any influences of the metal-semiconductor interface [114]. Finally, a tight diffusion barrier encapsulation against environmental gases and ensuring long-term measurement stability is required to prevent performance degradation [77].

Sensor Design

In the following, the parallel, site selective integration of large-diameter (> 2 nm) SGS-SWNTs by dielectrophoretic directed assembly into large channel (~ 1 µm), low-resistance pressure sensor devices is introduced. Figure 5.1 shows a schematic cross-section of the SWNT based pressure sensor composed of a silica/alumina membrane, encapsulating the carbon nanotube, which can be modulated by a top gate.

5.2 Experiments

Fabrication Process

Fabrication was initiated with a 200 µm thick doped silicon (Si) substrate, upon which 70 nm of silicon oxide (SiO_2) was grown at 1000°C by dry oxidation. Electrodes were structured by photolithography, evaporation of 5 nm Ti and 35 nm Pd, followed by lift-off. A 5 µl droplet of aqueous SWNT solution was then dispensed on the 6 × 6 mm chip for the dielectrophoretic nanotube assembly, and a sinusoidal potential of 0.5 V_p at 100 MHz applied to the contacted bias electrode and grounded substrate by a high frequency, high power signal generator (Agilent

5.2 Experiments

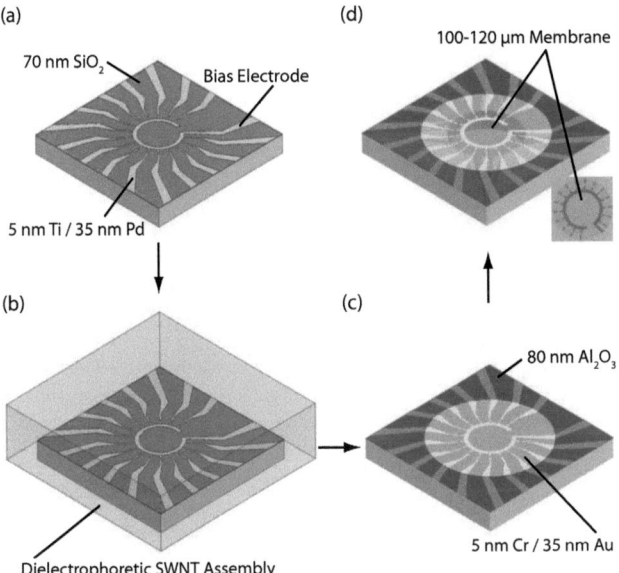

Figure 5.2: Process flow illustrating the parallel integration of SWNT based pressure sensors. (a) Radially arranged Pd electrodes on silicon oxide with gap sizes in the range of 1 μm are structured by photolithography, evaporation, and lift-off. (b) The chips are immersed in an aqueous SWNT dispersion for the selective dielectrophoretic deposition of SGS-SWNTs. The sinusoidal potential difference is applied to the bias electrode, while the other electrodes remain floating and are capacitively coupled to the grounded doped silicon substrate. (c) After integration, the SWNTs are covered by a protecting Al_2O_3 layer and a top gate is patterned above for CNFET gate modulation. (d) In the last step, the tri-layer silica/alumina/gold $100 - 120$ μm diameter and 190 nm thick membrane is released in a dry etching process after backside alignment to allow the SWNTs to be located on the membrane edges.

5 Piezoresistive Pressure Sensors with Parallel Integration of Individual SWNTs

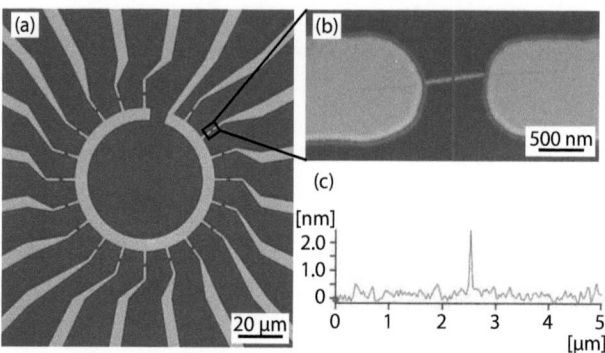

Figure 5.3: Circularly arranged electrodes and dielectrophoretic SWNT integration. (a) Optical microscopy image after lift-off. 19 individually accessible electrodes are patterned. The interior electrode is the common bias electrode for the capacitively coupled dielectrophoretic integration. (b) SEM image of a successfully bridged electrode gap showing the good SWNT alignment in order to minimize strain components other than from the main orientation axis. (c) AFM height scan in attractive tapping mode to determine the SWNT diameter, in this particular case slightly above 2 nm.

N5181A) [34]. After 60 s, the generator was switched off and the chip rinsed in deionized water (DIW). Scanning electron microscope (SEM) inspection and atomic force microscope (AFM) characterization of the deposited SWNTs was undertaken after this step. To improve contact adhesion, rapid thermal annealing at 400°C for 1 min was performed in a nitrogen (N_2) environment. The subsequent process steps are analogous to those used in previous studies [77, 105]. In short, 80 nm of Al_2O_3 was conformally grown on the chip surface to encapsulate the SWNTs in an ALD reactor (Picosun Sunale R-150B) at 300°C under a N_2 environment. The patterning of an unstructured top gate followed, by evaporating 5 nm Cr and 35 nm Au onto the high k alumina dielectric. In a last step, the backside structuring of the 120 μm diameter membrane was carried out by photolithography using infrared alignment. After resist development and an O_2 plasma cleaning step for 120 s at 200 W, the silica/alumina hard mask was etched in 6% buffered HF solution for 200 s. Finally the SiO_2/Al_2O_3/Au tri-layer membrane with a total thickness of 190 nm is released by inductively coupled plasma deep reactive ion etching (ICP-DRIE) employing the Bosch process. Complete membrane release is verified by optical microscopy, before the chip is packaged to a chip carrier and wirebonded. Figure 5.2 shows a schematic of the fabrication process flow.

5.2 Experiments

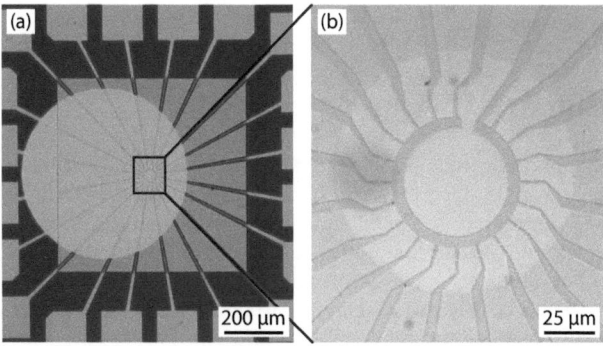

Figure 5.4: Processing images of the dielectrophoretically integrated SWNT pressure sensor. (a) Optical microscopy image depicting the protective rectangular Al_2O_3 layer and top gate covering the entire region to be located on the membrane, before the last etching step. On the edges, the bond pads of the chip are visible. (b) Optical microscopy image of the final device, showing the completely released 190 nm thick and 120 µm wide tri-layer $SiO_2/Al_2O_3/Au$ membrane. Partial transparency of the thin layers allows the electrodes contacting the SWNTs to be seen.

Figure 5.3 depicts the circularly arranged electrodes, to be located on membrane edges, ultimately the positions of highest strain. The electrode gap is imaged by SEM and the successful, aligned dielectrophoretic SWNT deposition is clearly identifiable. SWNT alignment is necessary to minimize strain components other than from the main orientation axis and achieved in the process. Gap sizes are consistently in the range of 800 nm. AFM characterization is necessary to determine the diameter of the deposited carbon nanotubes and confirms the large diameter of the synthesized SWNTs [48, 73]. It is noteworthy that the SWNTs are never exposed to any processing steps apart from the aqueous dispersion in surfactant stabilized solution and aluminum oxide conformal ALD encapsulation.

<small>Circular Electrodes</small>

In Figure 5.4, the protective rectangular Al_2O_3 layer is visible, which is deposited after annealing the SWNT to improve contact properties. Encapsulation not only provides protection against environmental influences but also ensures long-time stability. The circular top gate is evaporated over the central region of the chip and covers the entire region to be located on the membrane. The close up of Figure 5.4 depicts a completely released membrane with a diameter of 120 µm, where no notching is observed. This step concludes the pressure sensor fabrication. The thin metal layers are partially transparent, allowing the distinction between the different

<small>SWNT Encapsulation</small>

5 Piezoresistive Pressure Sensors with Parallel Integration of Individual SWNTs

layers. Only parallel processes are employed in the pressure sensor fabrication, thus enabling facile scale-up.

5.3 Results

DEP Yield

After dielectrophoretic integration, on the average 4-6 of the 19 separately accessible electrode gaps were bridged by an individual carbon nanotube, which is equivalent to a $20-30\%$ deposition yield and similar to previous studies [34]. For the unsuccessful deposition cases, no nanotubes at all were observed, the SWNTs were only attached to one electrode, or multiple distinguishable individual SWNTs bridged the gaps. As however only one functional carbon nanotube on the finalized pressure sensor is required, this deposition yield is sufficient. Further, the deposited SWNTs were in general straight, thus minimizing non principal strain components and local band gap opening by bending [67]. Figure 5.3 shows the successful dielectrophoretic integration of an individual SWNT on the pressure sensor electrode design.

SGS-SWNT Ratio

A 90% electrical characterization yield was achieved on all probed SWNTs. The robustness of the integration process is thus evidenced. Of all integrated SWNTs on average 3 out of 4 (75%) showed small bad gap semiconducting (SGS) behavior. An even higher ratio of SGS-SWNTs would be feasible when moving towards even higher dielectrophoretic deposition frequencies or employing solutions with different conductivity ratios [48].

Electrical Properties

The electrical transport behavior of a SGS-SWNT is presented in Figure 5.5. Top gate modulation makes the CNFET behavior apparent. When sweeping the gate potential between $V_g = \pm 5$ V a device resistance of in the range of $R = 50-80$ kΩ is recorded for a source-drain potential of $V_{sd} = 50$ mV. Contact barriers induce a difference between the source-drain current I_{sd} in the n-type and p-type on-state of the transistor, which leads to an asymmetry that may marginally influence the location of I_{min} in the off-state. Continuous gate sweeps with smaller gate modulation reduce hysteresis in the measurement, as fewer charges are trapped. Alternatively, pulsed gate measurements may be used [77]. The off-state of the SGS CNFET is determined by small gate voltage modulation to occur at $V_g = 0.5$ V, as can be seen in the inset of Figure 5.5. This is the position where the pressure sensor characterization is carried out because its resolution is maximized.

Electromechanical Properties

Figure 5.6 shows time dependent pressure sensor measurements at an applied gate voltage of $V_g = 0.5$ V and source-drain potential of $V_{sd} = 50$ mV. The upper curve displays the measured current output I_{sd} against the applied pressure p in the lower curve. A slight overshoot of the pressure in the chamber is detected due to the quick ramp rates. Fast current responses in the carbon nanotube transducer are however identified that way, including the overshoots when large enough. A

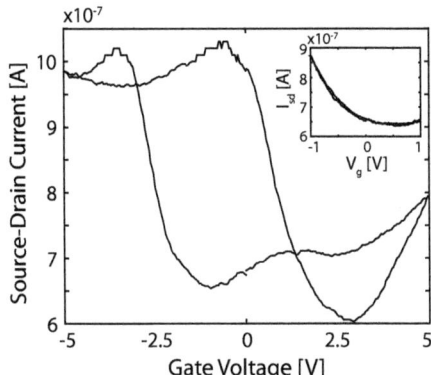

Figure 5.5: CNFET behavior of a SGS-SWNT driven at a source-drain potential of $V_{sd} = 50$ mV. A top gate is used to modulate the output. The off-state of the SGS CNFET is at $V_g = 0.5$ V, determined by small gate voltage modulation shown in the inset. Continuous gate sweeps with smaller gate modulation are used to reduce hysteresis in the measurement and determine the position of the off-state. The off-state is the location where the pressure sensor characterization is carried out because resolution is maximized.

sensor sensitivity of $S_0 \sim 0.25$ $\Delta R/R$/bar is extracted from the measurements with a resolution better than 50 mbar at an extraordinary low power consumption of less than 40 nW. The SWNT exhibited negative piezoresistive behavior, decreasing its resistance at increasing strain.

On the same membrane, 2 additional SGS-SWNTs were characterized with respect to their strain behavior. Both evidenced higher device resistances, lower sensitivities and reduced resolution, but also negative piezoresistive behavior. All probed SGS-SWNTs exhibited a piezoresistive response and long-term operation of the sensors over a period of 3 months was confirmed. Table 5.1 shows a summary of the electromechanically characterized SWNTs.

Additional Measurements

5.4 Discussion and Conclusions

Measurement observations are consistent with previous reports [77, 105, 106]. The sensitivity and resolution of the pressure sensors is greatest in the off-state of SGS-SWNTs [105] and contact properties have a major influence on the electrical response

Literature

5 Piezoresistive Pressure Sensors with Parallel Integration of Individual SWNTs

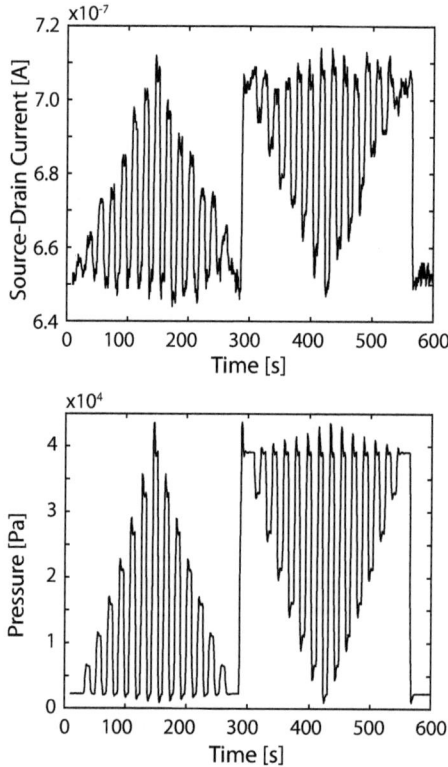

Figure 5.6: Time dependent pressure sensor measurements at $V_g = 0.5$ V and $V_{sd} = 50$ mV. The upper curve shows the measured current output I_{sd} against the applied pressure p in the lower curve. A sensor sensitivity of $S_0 \sim 0.25$ $\Delta R/R$/bar is extracted with a resolution better than 50 mbar at a power consumption of less than 40 nW. The SWNT showed negative piezoresistive behavior, decreasing its resistance at increasing strain.

5.4 Discussion and Conclusions

	Sensitivity	Off-state gate voltage	Resistance at off-state	Resolution
SWNT #1	0.25 ΔR/R/bar	0.5 V	74 kΩ	50 mbar
SWNT #2	0.12 ΔR/R/bar	-1 V	175 kΩ	100 mbar
SWNT #3	0.10 ΔR/R/bar	4.5 V	97 kΩ	100 mbar

Table 5.1: Overview of the electromechanically characterized SWNTs.

of the pressure sensor [98]. As the resistance increases, the sensitivity of the transducer elements tends to decrease, a prediction from the model of thermally activated transport. For a high resolution SWNT based pressure sensor, it is therefore essential to minimize contact resistances and increase transmission of the carbon nanotubes. This approach potentially provides the largest leverage for enhanced device performance. Top side metallization of the SWNTs contacts is one option for improving the contact properties and reducing device resistances. Surface interaction and channel lengths, however, may help explain the inconclusive results at higher device resistances.

DEP Solution processing of SWNTs is suggested to induce relatively few defects after dielectrophoretic integration, endorsing the potential use of these high-quality components in sensor and nanoelectronic applications [34,115]. The common availability of high-grade monodisperse SGS carbon nanotube solutions is expected to additionally improve the device integration yield [56]. The additional benefit of the introduced method is that multiple carbon nanotubes can be assembled onto one μ-sized membrane and that the best performing specimens can then subsequently be selectively combined together in a Wheatstone bridge configuration, for example, to cancel out temperature cross-sensitivity. Therefore, a comparatively low integration yield is sufficient for viable sensor assembly, without causing a substantial increase in fabrication costs due to the exclusively parallel assembly methods.

Scale-Up The scale-up of the introduced fabrication process is straight forward. Because of the purely parallel processes involved, especially for the individual SGS carbon nanotube integration, it is expected that large-scale fabrication entails substantial cost benefits. By additionally replacing some of the more expensive dry etching methods with solution phase etching processes, for example, processing costs could be further reduced. Overall, a high industrial and commercial potential is perceived.

Membrane The membrane design in the presented configuration is crucial, since the membrane has to be initially flat, preferably with a small effective initial tensile stress to avoid buckling. The SWNT encapsulating Al_2O_3 layer therefore not only has the role of passivating the carbon nanotubes, but also of compensating the compressive stress of SiO_2 through its tensile properties. A thickness ratio of 1 is determined to satisfy

5 Piezoresistive Pressure Sensors with Parallel Integration of Individual SWNTs

this requirement [114]. By slightly increasing the alumina thickness, an additional safety margin is added. In this study, pressure differences of up to 0.5 bar were easily withstood. Sensor sensitivity in a specific and well defined pressure range may be enhanced by reducing the membrane thickness and consequently stiffness, to allow higher strain rates at lower differential pressure differences to occur. An exact design approach requires a detailed membrane characterization to be carried out by finite element methods (FEM), evaluating the membrane thickness and diameter influence on the induced strain. This membrane characterization could then also be used for a detailed gauge factor analysis of the piezoresistive SWNT properties.

Conclusions   In conclusion, a scalable parallel assembly process for the fabrication of single-walled carbon nanotube based piezoresistive pressure sensors is demonstrated. The sensors show high sensitivity at a small size while consuming only minute power. Highest resolution is achieved for large-diameter, low-resistance devices at the off-state of small band gap SWNTs, as predicted by the model of thermally activated transport in carbon nanotubes. The directed assembly of SGS-SWNTs, enabled by their selective dielectrophoretic integration, and subsequent device processing yields robust and long-term stable sensors. All these elements demonstrate the robustness of the parallel small membrane pressure sensor fabrication process and reliability of the SWNT transducer element. It is expected that large-scale fabrication entails substantial cost benefits, which may ultimately result in commercially viable products. The repeatable assembly of high sensitivity ultra small pressure sensors, however, remains a challenge.

6 High-Yield Dielectrophoretic Assembly of Two-Dimensional Graphene Nanostructures

Parts of this chapter are published in:

B. R. Burg, F. Lütolf, J. Schneider, N. C. Schirmer, T. Schwamb & D. Poulikakos. High-yield dielectrophoretic assembly of two-dimensional graphene nanostructures. *Applied Physics Letters* **94**, 053110 (2009).

Abstract

Graphene handling is still dominated by serial mechanical exfoliation, which may well facilitate measurements in a laboratory environment but does not allow reliable larger-scale integration. Herein we demonstrate the controlled, high-yield (>90%), site-selective deposition of ultrathin few-layer (three to ten) graphene oxide by dielectrophoresis between prefabricated electrodes. Individual layers are found near the edges. Initially insulating, thermal reduction at 450 °C thins out the two-dimensional few-atom thick films and dramatically reduces electrical resistances down to 40 kΩ. Conductivities between 15 and 36 S/cm are obtained. The introduced method permits the nonintrusive, parallel, large-scale assembly of soluble two-dimensional nanostructures and sheets.

6.1 Introduction

The rise of two-dimensional graphene sheets [24] has led to the conception and development of unanticipated technical applications [116–118] and is recognized to have considerable potential as a building block in future nanoscale metallic electronics [25]. A major drawback however in advanced processing technology is the absence of parallel, controllable assembly methods. The prospect of bridging multiple contacted electrode gaps nonintrusively on a single device by purely parallel processes is therefore both very important and alluring. The research reported herein

Graphene Handling

makes a significant step in this direction. Employing dielectrophoresis, the large-scale high-yield integration and assembly of soluble ultrathin two-dimensional structures, such as (functionalized) graphene dispersed in (aqueous) solvents [119–123], is demonstrated.

DEP Dielectrophoresis allows the selective deposition or directed movement of spherical and one-dimensional micro- and nanoscale objects in nonuniform electric fields. This includes polymer particles [11], cells [124], DNA [125], nanowires [31], and bundles [60,126] or individual carbon nanotubes [30] for physical properties characterization [62]. The dielectrophoretic force $\mathbf{F}_{DEP} = (\mathbf{p} \cdot \nabla)\,\mathbf{E}$ is exerted on an induced dipole moment $\mathbf{p}$ of a polarizable particle in a nonuniform electric field $\mathbf{E}$, which is proportional to the electric field gradients. Particles move toward regions of high electric field strength if the polarizability of the particles is greater than that of the suspending medium; otherwise they are repelled from those regions [12–14].

Capacitive Coupling The controlled deposition of individually accessible devices at very high integration densities is achieved by capacitively coupling the metallic leads to a conductive substrate. The self-limiting single-walled carbon nanotube (SWNT) assembly between prefabricated electrodes [32], as well as the separation of metallic from semiconducting SWNTs [35], is reported under specific deposition conditions. The accumulation of μ-thick graphite oxide soot films by dielectrophoresis has further been observed [127].

6.2 Experiments

Chip Fabrication The electrodes of the present study are prepared by standard photolithography (Figure 6.1). A single mask of image reversal photoresist (AZ 5214E) is used to dry etch embedded electrode pockets, followed by palladium sputtering and lift-off. Between the 40 nm thick semicircular metallic electrodes, gaps of $1-2$ μm are created. An insulating thermal dry oxide layer ($t_{ox} = 285$ nm) allows the capacitive coupling between the p-type doped Si substrate and the prepared electrodes. Capacities are in the range of $C \approx 10$ pF. The chips are subjected to a thorough and aggressive cleansing process in solvents and oxygen plasma before use. This results in extremely clean, flat, and smooth substrate surfaces.

Wiring On one chip, ten positions are available for dielectrophoretic deposition. In all experiments one counter electrode is additionally grounded (direct coupling), thus keeping the electric field in the gap constant, even after deposition, to investigate the influence on the deposition of ultrathin film nanostructures. All other electrodes remain floating (capacitive coupling).

Solution Preparation Stable aqueous dispersions of exfoliated graphene oxide (GO) are synthesized by a modified Hummers method [128], in which a very long acid oxidation period is

6.2 Experiments

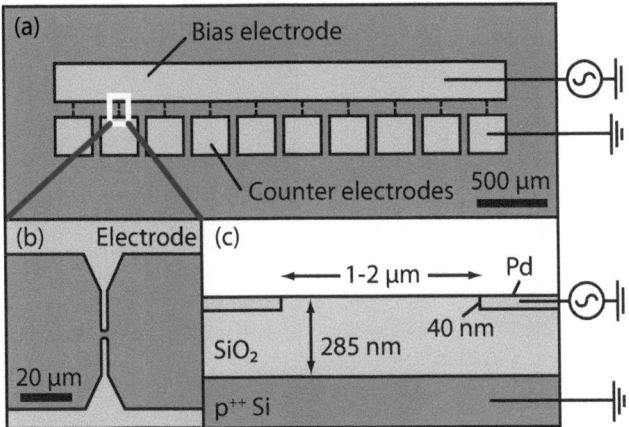

Figure 6.1: Chip schematic. (a) illustrates the bias electrode used for applying the bias potential from the sinusoidal function generator and the counter electrodes, which are capacitively coupled to the conductive substrate except for one, which is directly coupled to the ground. Inset (b) shows in detail the finger electrodes in yellow, over which the two-dimensional graphene nanostructures will bridge. The cross section of the finger electrode gap is depicted in (c).

combined with a thorough purification process to obtain highly oxidized, exfoliated, and extremely pure GO dispersions [129,130]. In short, 1 g of natural graphite flake (Fluka, purity >99%, particle size of < 20μm) is mixed for 5 days in a 62.1 ml solution of concentrated H_2SO_4 containing 0.75 g $NaNO_3$ and 4.5 g $KMnO_4$. It is then washed in 100 ml 5 wt % H_2SO_4 and reacted with 3 g of 30 wt % H_2O_2 in H_2O. Impurities are removed by multiple (15×) centrifugation and ultrasonic precipitate resuspension in aqueous 3 wt % H_2SO_4 and 0.5 wt % H_2O_2. After repeating the procedure three more times with H_2O and letting the dispersion rest for 1 day, a last centrifugation cycle (8×) in H_2O is performed with the supernatant half of the solution. After drying the single-layer particles, they are ultrasonically redispersed in water at a concentration of 1 mg/ml.

A droplet (5 μl) of the aqueous GO solution is dispensed on the chip, and a sinusoidal potential difference is applied to the bias electrode (typically 5 V_p at 5 MHz) by an sinusoidal function generator (LeCroy LW420B). The conductive, capacitively coupled substrate remains grounded during the entire procedure. The generated electric field reaches average values in the range of $E_{rms} \approx 10^6 - 10^7$ V/m between the capacitively coupled electrodes and is highly localized in the vicinity of the elec-

Graphene DEP

6 High-Yield Dielectrophoretic Assembly of Two-Dimensional Graphene Nanostructures

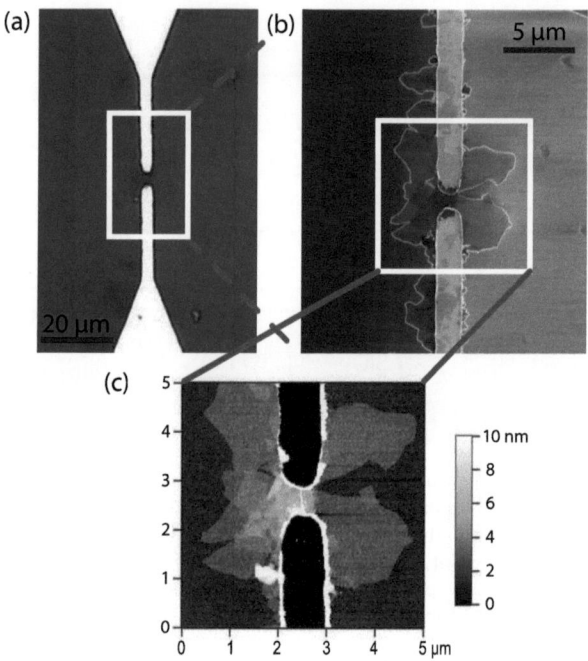

Figure 6.2: Dielectrophoretic FLGO deposition. (a) Light microscopy image showing the palladium electrodes and revealing the accumulation of FLGO in the regions of highest field gradients. (b) SEM image of the same position with clearly visible contours of the flakes. (c) AFM image of the identical location, showing overlapping layers, twists, and creases of the thin film. Individual GO layers are found near the edges of the electrodes.

trode gaps. The GO sheets contained in the solution do not influence the makeup of the electric field due to their dilute concentrations. After 60 s the generator is switched off, the droplet is blown off by a stream of nitrogen gas, and the is chip rinsed in de-ionized water. The sample is characterized by light microscopy, scanning electron microscopy (SEM), atomic force microscopy (AFM), and electric transport measurements.

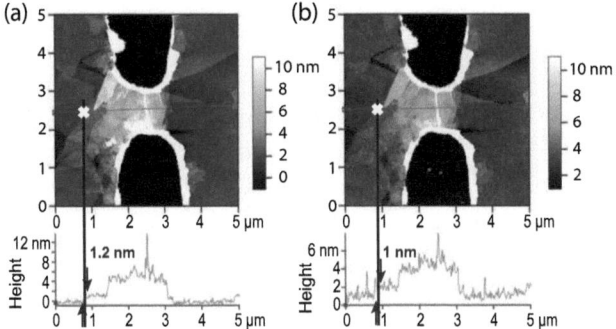

Figure 6.3: Consecutive AFM scans of the position depicted in Figure 6.2 after (a) dielectrophoretic deposition and (b) thermal annealing at 450 °C. The reduction in layer thickness is visible in the height profiles. Creases and folds are overproportionally thinned during the reduction process. Note the different scale bars in the height profiles.

6.3 Results

Few-layer GO (FLGO) dielectrophoretic deposition between the metallic electrodes is observed in nine or ten positions for all performed experiments, resulting in a larger than 90% yield. An example of a bridged gap is shown in Figure 6.2. Metallic graphene sheets obtained by thermal reduction from FLGO are reported to have higher mobility and conductance than those obtained by chemical reduction, employing hydrazine, for example [131, 132]. It is assumed that the thermal reduction route is more complete compared to the known chemical methods. When heated up to ≈1200 °C, oxidized functional groups are reduced and graphene sheets are recovered.

FLGO Deposition

Thermal annealing of the samples is performed in a rapid thermal anneal oven (J.I.P. Elec JetFirst 100) in an inert N_2 environment. Slow heating ramps of 20 °C/min are used to prevent film gasification, ripping, and exfoliation due to rapidly expanding H_2O adsorbed between the FLGO layers [131]. Thermal graphenization of FLGO is performed at 450 °C for 1 h. Above these temperatures the palladium contacts disintegrate.

Thermal Reduction

Reduction in the FLGO thin films is confirmed by a sheet conductivity increase and decrease in layer thickness. Initial average FLGO layer thicknesses lie between 3 and 10 nm at their thickest area between the electrodes and are thinned to 2 − 8 nm after reduction, with creases and folds being overproportionally thinned during the process. This reveals a reduction in the interstitial water between the individual

FLGO Layer Thickness

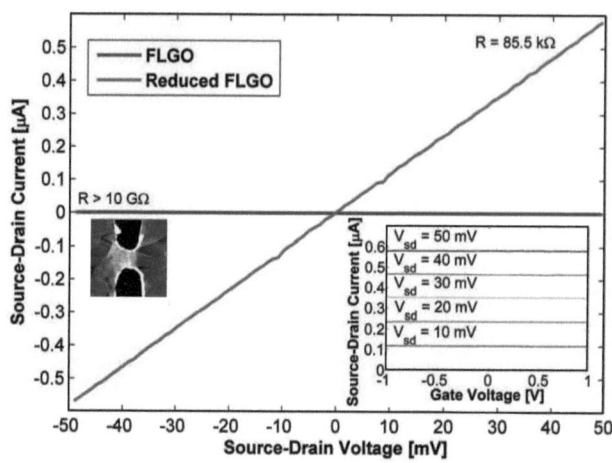

Figure 6.4: Current-voltage characteristics of the FLGO and thermally reduced FLGO junction introduced in Figures 6.2 and 6.3, displaying an insulating behavior before the thermal reduction and reaching an electrical resistance of 85.5 kΩ and conductivity of 23.5 S/cm after being heated up to 450 °C for 1 h in an inert N_2 environment. The inset shows no gate dependency.

FLGO layers and functional groups in the material.

Consecutive AFM scans on the same location are shown in Figure 6.3. The height steps of overlapping individual FLGO flakes decrease from ≈1.2 to ≈1.0 nm after thermal treatment. FLGO proves to be thermally stable and its morphology remains the same after thermal treatment.

FLGO Conductivity   Current-voltage (I-V) measurements at room temperature and ambient conditions, showing the transition from an insulating ($R > 10$ GΩ) to a conducting junction, are presented in Figure 6.4. The electrical resistance is decreased by over six orders of magnitude, with the final resistance after thermal reduction at 450 °C reaching values as low as $R = 40$ kΩ for all experiments. No further processing is required, such as top-side metallization, with the FLGO simply covering the contact electrodes. Good Ohmic contacts are verified by the linear relationships. No gate voltage dependency is observed, inferring metallic behavior of the sheets.

Sheet Conductivity   When approximating the length to width ratio (L/w) of the flake 1:1, the electrical sheet resistance R_s of the contact junction lies within the same range as the measured two point resistance. Sheet resistances for the presented data lie within 40 – 350 kΩ/ sq. Conductivity values σ average from 15 to 36 S/cm when taking the measured sheet thickness t into account. This is within the range of previously obtained values [132].

6.4 Discussion Conclusions

With the known repeat distance along well-hydrated GO layers being 11 Å [133], the number of deposited FLGO layers is estimated to lie between three and ten. Near the edges of the electrodes, individual GO layers are distinguished. When keeping the electric field and consequently the dielectrophoretic field force constant between the electrodes through direct coupling during the deposition process, thin film graphite oxide is deposited with thicknesses reaching 80 nm. This indicates that by carefully controlling the electric field build-up, configuration and strength, the amount and thickness of deposited two-dimensional sheets can be controlled.

Number of GO Layers

The introduced process allows the reliable, controlled, and parallel site-selective deposition of few-layer graphene flakes on photolithographically prefabricated structures. The investigation constitutes an important step toward the controlled deposition of ultrathin two-dimensional materials, such as (water) soluble and/or functionalized graphene [119–123]. Since these materials possess a higher conductivity and permittivity than GO due to their inherently more conductive nature, which causes a higher polarizability [14], the proven attractive dielectrophoretic force may increase in magnitude, even when the sheets do not possess a surface charge. The availability of highly conductive soluble pristine graphene is therefore anticipated to further establish the method, which may pave the way for unprecedented nanoscale material characterization and device fabrication approaches.

Conclusions

7 Dielectrophoretic Integration of Single- and Few-Layer Graphenes

Parts of this chapter are published in:

B. R. Burg, J. Schneider, S. Maurer, N. C. Schirmer & D. Poulikakos. Dielectrophoretic integration of single- and few-layer graphenes. *Journal of Applied Physics* **107**, 034302 (2010).

Abstract

The dielectrophoretic integration of single- and few-layered graphenes from three distinct graphene suspensions is presented, enabling the parallel assembly of individual two-dimensional nanostructures at predefined locations. The first suspension is an aqueous solution of graphene oxide, the second is ultrasonically exfoliated pristine graphene in N-methyl-pyrrolidone (NMP), and the third is exfoliated graphene in surfactant-stabilized 1 wt % aqueous SDBS solutions. The most crucial aspect for the successful thin flake deposition is the solution quality of the exfoliated graphene. After dielectrophoresis, single-layer graphene oxide is placed between the electrodes, which, while initially insulating, recovers its electrical conductivity following thermal reduction. From the chemically unmodified graphene-NMP solutions, the directed assembly of electrically active few-layer graphene flakes is realized, with flake thicknesses in the range 8 – 30 nm. Liquid phase exfoliation in water-surfactant solutions yields significantly thicker flake dimensions from 50 to several 100 nm due to the higher enthalpy of mixing in the dispersion. To achieve single-layer pristine graphene dielectrophoretic deposition, higher solution qualities must be available, consisting largely of single-layer graphene sheets. The reported research provides an important framework for parallel fabrication approaches of graphene-based devices.

7.1 Introduction

A vast amount of research has been spurred in recent years, devoted to the development of methods for preparing individual pristine graphene sheets, a single layer of sp2-hybridized carbon atoms bound together in a hexagonal lattice [24]. The high

Graphene Applications

7 Dielectrophoretic Integration of Single- and Few-Layer Graphenes

promise of potential applications, such as graphene-based electronics or solid-state gas sensors, has largely driven this rise of graphene [25]. Furthermore, chemically modified graphene, such as graphene oxide (GO), has also emerged as a promising candidate for components in applications [116]. These include chemical detectors at the molecular level, energy-storage materials, paperlike materials, polymer composites, liquid crystal devices, mechanical resonators, and highly conductive transparent electrodes [134].

Graphene Preparation
Two distinct approaches exist for preparing these two dimensional structures. One is to mechanically cleave layered graphite into individual planes and the alternative route is to epitaxially grow graphitic layers on top of other crystals [27]. A scalable approach, offering the possibility of high-volume production and well suited for, but not limited to chemical functionalization, is the ultrasonic cleavage of graphite to form colloidal graphene suspensions. This leads to stable solutions of graphene crystallites in varying concentrations and different solvents [116]. The lack of subsequent parallel, reliable, and controllable assembly and integration techniques, however, still hinders progress toward the large-scale application of graphene-based devices.

Graphene DEP
Out of solution dielectrophoretic deposition assembly provides this prospect by attracting dispersed particles to prefabricated electrodes through electric fields [11]. This technique has been recently refined and improved, for carbon nanotube integration among others [30, 62]. Crucial to a high integration yield is the quality of the prepared solution, which must consist of stable, homogeneously dispersed nanostructures in sufficiently high concentration [34].

Graphene Solutions
A prerequisite for the dielectrophoretic graphene integration therefore is the availability of solutions made up of individually dispersed graphene sheets in the micrometer size range. This was first achieved for chemically modified GO aqueous solutions [119]. The electrical contacting of solution phase processed graphene is an additional requirement for the applicability of the method [120]. Earlier work has shown the dielectrophoretic integration of micrometer thick graphite oxide soot [127], as well as the deposition of few- and single-layer GO sheets between metallic electrodes [51, 89, 135]. More recently, the assembly of chemically unmodified few-layer graphene (FLG) was published [136].

Nomenclature
Concerning nomenclature, FLG refers to multilayered graphene up to about 10 layers, since the material reaches the limit of three dimensional graphite in terms of electrical and dielectric properties at that thickness [137].

Approach
In the present paper, three distinct graphene dispersions for the subsequent dielectrophoretic integration of single- and few-layered graphenes are presented, in order to evaluate their viability in general and disparity with respect to each other, respectively. The first involves chemically modified GO in aqueous solutions, the second ultrasonically exfoliated graphene in solvents whose surface energies match that of

graphene, and the third ultrasonically exfoliated graphene in surfactant-stabilized solutions. Based on the results of the dielectrophoretic integration, the quality and suitability of the colloidal solutions, prepared by fundamentally different dispersion principles, are investigated.

7.2 Experiments

Aqueous GO solutions are prepared by a modified Hummers method [128], where a long oxidation period is combined with a thorough purification process. In short, a mixture of sulfuric acid (H_2SO_4), sodium nitrate ($NaNO_3$), and potassium permanganate ($KMnO_4$) is used to oxidize the natural graphite powder. This yields highly oxidized, exfoliated, stable, and extremely pure solutions of chemically modified graphene and consist of the same dispersions as those used in earlier work [51]. The solution with a concentration of 0.25 mg/l remains stable for months, without any indication of sedimentation. GO Solutions

The exact structure of GO is still under debate but includes hydroxyl and epoxy functional groups in the plane and carbonyl and carboxyl groups at the sheet edges [138]. Due to the distortion of the graphitic network, GO sheets are electrically insulating. Oxygen content reduction, however, renders the sheets electrically conductive again. Chemical reduction is mostly performed in liquid phase environments of GO colloidal solutions utilized for subsequent deposition [139], whereas thermal reduction is better suited and less hazardous for already deposited GO flakes [140]. GO Structure

For this reason, thermal graphenization in an inert nitrogen (N_2) environment at 450 °C for 1 h is the method of choice to render dielectrophoretically deposited GO sheets electrically conductive. To prevent film gasification and ripping due to rapidly expanding trapped H_2O, slow heating ramps of 20 K/min are employed [131]. The maximum reduction temperature is limited by the thermal stability of the metallic electrodes. Thermal Reduction

Liquid phase production of chemically unmodified graphene is obtained by natural graphite dispersion in a low power ultrasonic bath (Bandelin Sonorex). Two distinct solvents are employed to facilitate the layer exfoliation and stabilize the obtained solutions. The first consists of N-methyl-pyrrolidone (NMP), whose surface energy is well matched to that of graphene, so that exfoliation freely occurs and steric stabilization takes place [123]. Chemically Unmodified Graphene
NMP

As this solvent is expensive and requires special care when handling, the alternate approach uses water-surfactant solutions, specifically 1 wt % sodium dodecylbenzenesulfonate (SDBS) [141]. The surfactant adsorbs onto the graphene flakes and imparts an effective charge, which is then reported to stabilize the graphene dispersion by the electrostatic repulsion of the surface coated graphene flakes, analogous SDBS to single-walled carbon nanotube solutions [54, 55]. Both solvents represent two fundamentally opposed dispersion principles. Reports on additional solvents for liquid phase graphene exfoliation are attributed to one of the described mechanisms above [142–144].

Solution Preparation
For the solution preparation, 1 g of large graphite flakes (NGS Naturgraphit GmbH, flake size of 5 – 10 mm) are dissolved in 100 ml of the respective solvent and washed by centrifugation to remove any residual impurity contamination (3×5 min at 10 000 rpm). The precipitate is then resuspended and subjected to ultrasonication for 15 min to cause exfoliation. The solution is finally centrifuged at 1 000 rpm for 60 min, the supernatant half decanted and retained. The comparatively low centrifugation speeds are required to avoid particle sedimentation, which would consequently yield too dilute graphene concentrations in the solutions as well as too small flake sizes [123]. To improve contact adhesion, the chips with the deposited flakes are annealed in a rapid thermal anneal oven (J.I.P. Elec JetFirst 100) under an inert N_2 environment at 200 °C for 1 min with a quick heating ramp of 10 K/s.

Chip Fabrication
Dielectrophoresis (DEP), which allows the site-selective deposition of micro- and nanoscale objects in nonuniform electric fields, is then carried out on chips containing 20 individually accessible electrodes. The gold electrodes have a thickness of 40 nm and a gap of 1 – 2 μm between them. They were fabricated using standard lithography and lift-off techniques [34].

Capacitive Coupling
Particles in a solution move toward regions of high electric field strength if the polarizability of the particles is greater than that of the suspending medium; otherwise they are repelled [14]. A thin oxide layer between the prefabricated metallic electrodes and highly doped silicon substrate allows for the capacitive coupling of the electrodes and consequently enables large-scale integration [31]. This is relevant for the scale-up of a large number of devices onto a single chip, which can then be integrated into a wide range of micro- and nanoelectromechanical systems and transducer applications.

DEP Process
An sinusoidal function generator (LeCroy LW420B) is connected to one side of the electrodes and creates a sinusoidal potential difference (0.7 – 6 V_p at 1 – 10 MHz). Then, a droplet of the colloidal graphene solution (3 – 5 μl) is dispensed onto the chip and the voltage source switched off after 60 s. Following deposition, the chips are rinsed in deionized water for all experiments performed with aqueous solutions and in successive baths of acetone, isopropanol, and de-ionized water for experiments carried out with NMP based solutions, in order to remove all remaining traces of the suspension. The chips are dried by a stream of nitrogen gas. No topside metallization is performed, with the graphene nanostructures simply adhering to the surface of the contact electrodes.

DEP Background
The dielectrophoretic force $\mathbf{F}_{DEP} = (\mathbf{p} \cdot \nabla) \mathbf{E} = v \mathrm{Re}\left[\tilde{\alpha}\right] (\mathbf{E} \cdot \nabla) \mathbf{E}$, where $\mathbf{p}$ is the effective dipole moment, v is the volume of the particle, and $\tilde{\alpha}$ is the complex effec-

tive polarizability, is highly dependent on the gradient of the nonuniform electric field and the polarizability of the particle [14]. The strongest external influence on the dielectrophoretic force therefore is the applied voltage, which must be adjusted according to the polarizability of the particles in the solution. Optimal deposition parameters are reached just above the deposition threshold, thus minimizing the induced forces. Additionally, the magnitude of the dielectrophoretic force is proportional to the volume of the particle, which is directly related to the flake thickness. Thicker sheets are consequently subjected to a larger attraction force.

Topographic, material, and electrical characterization of the samples is performed by light microscopy, scanning electron microscopy (SEM), atomic force microscopy (AFM), Raman spectroscopy, and electric transport measurements. Electrical measurements of the source-drain current I_{ds} versus gate voltage V_g are performed by continuous linear gate sweeps with V_g going from $-V_{g\,max}$ to $+V_{g\,max}$, with a constant sweep rate and constant source-drain voltage V_{ds}. I_{ds} is amplified by a low noise, large bandwidth current to voltage amplifier and acquired by a voltage meter.

Sample Characterization

7.3 Results

First, the results obtained by experiments performed on aqueous GO solutions are presented, followed by dispersions of graphene in NMP, and concluded with surfactant-stabilized aqueous graphene solutions.

AFM scans in attractive tapping mode are used to determine the thickness of the dielectrophoretically deposited flakes and enables conclusions to be drawn on the number of graphene layers lying between the electrodes. Figure 7.1 shows that it is possible to deposit single-layer GO by optimizing the dielectrophoretic deposition parameters (0.7 V_p at 10 MHz). Differential heights between the underlying silicon oxide substrate and flake surface are found to be in the range of $1.0 - 1.4$ nm. This is in agreement with values determined for well-hydrated GO monolayers and suggests that the sheets rehydrate on their surface under ambient conditions after partial thermal reduction [119, 133]. Single-layer GO can be visually distinguished in the SEM; therefore, AFM topography characterization was only performed on the most promising specimens. Overall DEP single-layer GO deposition yields bridging the electrodes for the employed parameters lie between 10% and 20%. For the unsuccessful deposition cases, no sheets at all are observed, the sheets are only attached to one electrode, or occasionally multiple distinguishable individual sheets bridge the gaps on top of each other. These results show that the dispersed GO aqueous solutions are largely made up of single-layer sheets.

GO Flake Thickness

As can be seen in Figure 7.1, the sheets are not perfectly flat, but may contain creases and folds and show irregular patterns at the edges, which is most probably

Single-Layer GO

7 Dielectrophoretic Integration of Single- and Few-Layer Graphenes

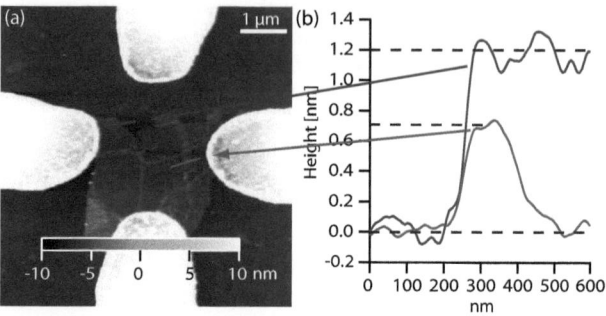

Figure 7.1: Thickness measurements of a single-layer GO flake. (a) AFM scan of a GO flake deposited by DEP over three electrodes. (b) AFM height traces at different sections of the flake. The blue trace shows a step height of 1.2 nm between the GO sheet and the underlying substrate, consisting of the thickness of a fully hydrated layer of GO. The red trace shows the differential thickness at a fold, made up of two overlapping GO sheets, which is around 0.7 nm. This suggests an inherent layer thickness of 0.35 nm, in the range of the graphite interlayer spacing of 0.34 nm.

due to the asperity at the electrode edges that cause local inhomogeneities in the electric field. The differential thickness at a fold, which is made up of two overlapping sheets, is measured to be around 0.7 nm. This implies an inherent flake thickness of 0.35 nm and is in the range of the graphite interlayer spacing of 0.34 nm. Consequently, no water is assumed to be intercalated between the folds after thermal annealing.

Thermal Reduction The incomplete reduction process of the deposited GO sheets is confirmed by the strong D-band peak in the Raman spectrum [119]. This is shown, for example, in Figure 7.2 for the flake imaged by SEM. The reference spectrum of few-layer pristine graphene on silicon oxide, prepared by mechanical exfoliation, is given as a comparison [145] and emphasizes the high defect density of the thermally reduced GO sheets.

GO Electrical Characterization Figure 7.3 presents the electrical characterization under ambient conditions of a narrow and flat dielectrophoretically deposited GO flake, without any creases or folds and a thickness slightly above 1 nm. Upon deposition, the GO monolayers are electrically insulating [51]. After annealing, the overall electrical resistance, consisting of the sheet and contact resistance, is $R = 128$ MΩ for the presented sample. Despite this very large value, the current-voltage characteristics show linear behavior up to a measured source-drain voltage of $V_{sd} = \pm 1$ V, which implies that no Schottky barriers and additional barriers exist at the contact interfaces. The obtained sheet

7.3 Results

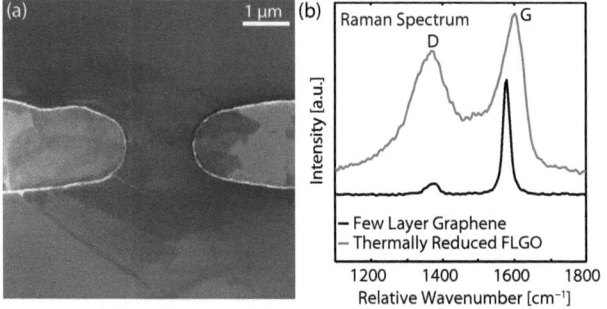

Figure 7.2: (a) SEM micrograph of a single GO flake and (b) its corresponding Raman spectrum after thermal reduction. The strong D-band peak confirms the incomplete reduction process and the high defect density. The Raman spectrum of a few-layer pristine graphene flake, fabricated by mechanical exfoliation, is given as a reference.

resistance of $R_S = 73$ MΩ/sq is much higher than for single-layer pristine graphene and FLG oxide [51] and corresponds to a conductivity value of $\sigma = 0.14$ S/cm. This low conductivity value is attributed to the fact that the sheets are still considerably oxidized, which is confirmed by the Raman spectrum. Slight p-type behavior of the GO sheets is observed for gate voltages in the range of $V_g = \pm 10$ V. This is in agreement with previous observations and is attributed to doping by oxygen and/or water absorption [146]. Also sheet doping, originating from the lateral diffusion of metal atoms at the contacts along the GO sheet during the annealing process, is possible [147].

Qualitatively and quantitatively, the dielectrophoretically deposited and thermally reduced sheets are comparable in their characteristics to randomly dispensed single-layer GO sheets on a surface, which are then subsequently thermally annealed and photolithographically contacted by sets of metallic electrodes [131,140,146,147]. This evidences that the introduced method constitutes a considerable step toward the parallel integration of micrometer sized monolayer GO sheets, with no serial steps required anymore.

Parallel Integration

Dielectrophoretic deposition results of liquid phase exfoliated graphene in NMP solutions are presented in Figure 7.4. The optimal deposition parameters differ substantially with respect to those used for aqueous GO solutions and much larger electric field strengths are required to obtain a deposition yield of 50%-60% (6 V_p at 1 MHz). For the unsuccessful deposition cases, no sheets at all are observed and it was never seen that individual flakes deposit on top of each other. AFM scans show that it is possible to deposit very thin pristine graphene flakes with thicknesses

NMP Sheets

7 Dielectrophoretic Integration of Single- and Few-Layer Graphenes

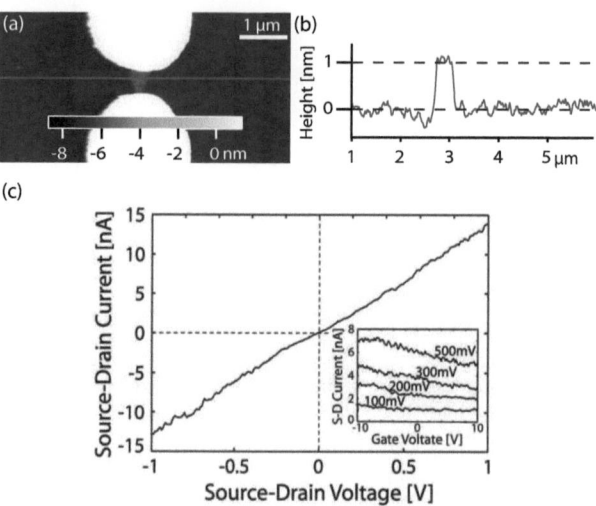

Figure 7.3: Electrical transport measurements on a single-layer flat GO sheet. (a) AFM scan of the electrically probed sample. (b) AFM height trace of the sample, displaying a flake thickness of a little more than 1 nm. (c) Current-voltage characteristics of the GO layer, exhibiting an electrical resistance of $R = 128$ MΩ and conductivity of $\sigma = 0.14$ S/cm after a thermal reduction at 450 °C for 1 h in an inert N_2 environment. The inset shows a slight p-type behavior of the reduced GO sheet.

reaching down to 8 nm. This coincides with the limit where few-layer chemically unmodified graphene flakes can be considered. On average, flakes with thicknesses of 10 − 30 nm are observed for the introduced system.

NMP Solution Quality   As the dielectrophoretic attraction force increases proportionally to the volume of the flake [14], this allows the conclusion to be drawn, that the NMP solutions are made up of graphene sheets with a maximal thickness comparable to that of the deposited flakes. Single-layer graphene may be present in the solution but not in sufficiently high concentrations to be successfully dielectrophoretically attracted to the electrodes, without having to compete against larger flakes subjected to a stronger attraction force. Graphene solutions largely consisting of single-layer graphene are therefore expected to yield devices of the same, which underpins the importance of the solution quality.

NMP Measurements   Electric transport measurements on these chemically untreated flakes naturally result in much higher conductivity values than for the previously discussed GO

7.3 Results

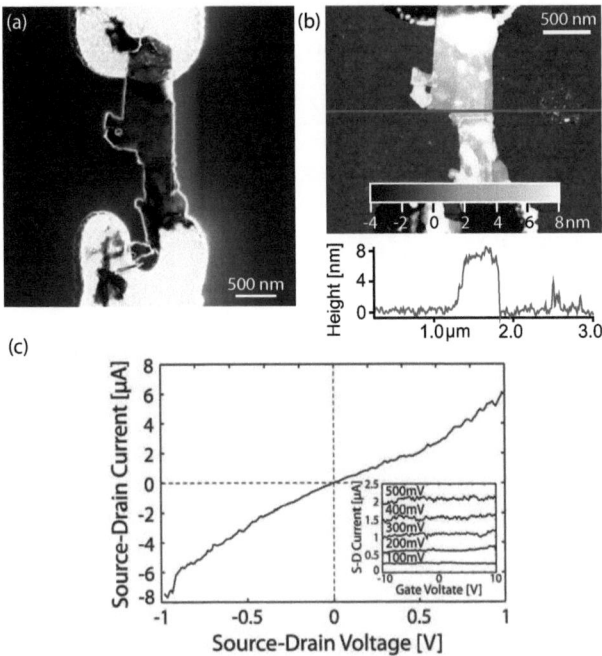

Figure 7.4: Dielectrophoretically deposited FLG from NMP solution. (a) SEM image of an electrode pair bridged by FLG. (b) AFM scan including height trace, displaying a flake thickness of 8 nm. (c) Current-voltage characteristics of the FLG flake, exhibiting an electrical resistance of $R = 270$ kΩ and conductivity of $\sigma = 15.43$ S/cm after thermal treatment at 200 °C for 1 min in an inert N_2 environment. The inset shows no gate dependency.

flakes. A representative resistance, evidenced by the flake shown in Figure 7.4, is $R = 270$ kΩ, which corresponds to a conductivity of $\sigma = 15.43$ S/cm when taking its geometry into account. This is two orders of magnitude higher than for single-layer GO and comparable to previously reported values without top-side metallization [136]. The current-voltage characteristics are not perfectly linear, which presumably arise due to contact barriers at the interface of the thin sheet and electrodes. These are most probably hidden in the GO case due to the much larger inherent sheet resistances. The metallic behavior is evidenced by the absence of a gate dependency.

83

7 Dielectrophoretic Integration of Single- and Few-Layer Graphenes

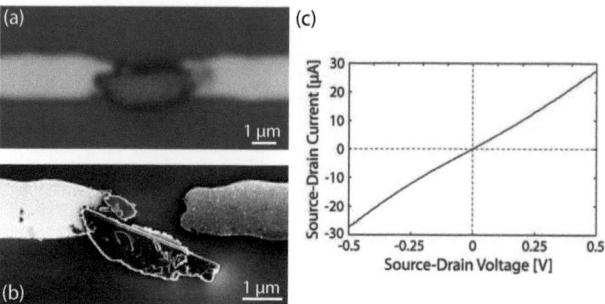

Figure 7.5: Dielectrophoretically deposited thick graphite flake from aqueous SDBS solution. (a) Light microscopy image of an electrode pair bridged by a several hundred nanometer thick graphite flake. (b) SEM image after flake dislocation with a micromanipulator. (c) Current-voltage characteristics of the bridged flake, exhibiting an electrical resistance of $R = 20$ kΩ and conductivity of $\sigma = 1.50$ S/cm after thermal treatment at 200 °C for 1 min in an inert N_2 environment.

SDBS Flakes Finally, the results of liquid phase exfoliated graphene in aqueous SDBS solutions are analyzed in Figure 7.5. With identical dielectrophoretic deposition parameters as used for NMP solutions (6 V_p at 1 MHz), slightly lower integration yields of 40% − 50% are achieved. Aqueous SDBS graphene solutions exhibit much thicker flakes than NMP solutions though, from an observed minimum of around 50 nm to several hundred nanometers, as depicted in Figure 5. This implies an extremely low yield of single-layer and FLG exfoliation during the solution preparation and the remaining of thick graphite particles in the solution. The exact thickness of these thick flakes is difficult to determine by AFM, due to their height. Furthermore, they are easily dislocated by a micromanipulator, as can be seen in the SEM image, or even detached by ultrasonic treatment. This manifests their low van der Waals adhesion forces to the underlying electrode, an observation never made for flake thicknesses below 50 nm. The high flake volume and large contact area on the electrode lead to low resistance values of $R = 20$ kΩ in their metallic behavior. Assuming a flake thickness of $t = 500$ nm, this leads to a conductivity of $\sigma = 1.50$ S/cm, one order of magnitude lower than for thin flakes dispersed in NMP. This is due to the much more pronounced flake thickness, originating from the low exfoliation yield in the surfactant-stabilized graphene solution, as the current has to pass through multiple graphitic layers.

7.4 Discussion and Conclusions

The ability to dielectrophoretically deposit single-layer GO and pristine graphene is solely dependent on the quality of the solution, as the dielectrophoretic force increases proportionally to the flake thickness. As shown for the aqueous GO solutions, once it is possible to fully exfoliate natural graphite in liquid phase without chemical modification, the integration of pristine graphene layers will be possible. This requires significant progress in the field of liquid phase graphene exfoliation to be made. Alternatively, a complete reduction in deposited GO sheets on metallic electrodes could lead to the same result, which, however, is limited by the thermal stability of the electrodes for thermal reduction processes. In any case, a high quality solution of fully exfoliated and dispersed pristine graphene sheets is of fundamental and practical importance for the successful dielectrophoretic integration of single-layer graphenes, as shown for homogeneously dispersed single-walled carbon nanotube solutions [34]. *(Solution Quality)*

Comparing the deposition results of pristine graphene flakes dispersed in NMP and aqueous SDBS solutions, it is confirmed that liquid phase exfoliation of graphene works best for solvents whose surface energy is well matched to that of graphene, so that exfoliation occurs freely, in contrast to surfactant-stabilized solutions [141]. The average deposited flake thickness from NMP solutions ranges from below 10 to 30 nm. For aqueous SDBS solutions, the range goes from 50 to several 100 nm. This infers that surfactants may be well suited for solution stabilization but do not facilitate ultrasonic graphene exfoliation. *(NMP vs. SDBS)*

The substantially lower potentials required between the electrodes for the successful integration of single-layer GO sheets are explained by the hydrophilic nature of GO. The hydroxyl and epoxy functional groups in the plane and carbonyl and carboxyl groups at the sheet edges partially dissociate in water and give GO a negative surface charge. This induces a surface conductivity and consequently a higher polarizability of the suspended structure. The Clausius-Mossotti factor, which describes the relationship between the dielectric constants of two different media and is a key coefficient in the dielectrophoretic force formulation, accordingly increases in magnitude [14]. This elucidates the comparatively facile dielectrophoretic integrability of the intrinsically insulating material at values which do not work for the other graphene solutions. Once deposited, the GO surface conductivity in the aqueous environment balances the surface potential between the bridged electrodes, so that the capacitively induced electric field is significantly reduced and no further flakes are deposited on top of previously integrated flakes, as long as the solution concentration and applied potential are not exceedingly large. The same self-limiting deposition argument holds for the inherently conductive, chemically unmodified FLG and graphitic flakes. The much higher conductivity for these materials further *(GO Surface Conductivity)*

7 Dielectrophoretic Integration of Single- and Few-Layer Graphenes

Flake Robustness
explains why pristine flakes are never seen to deposit on top of each other.
A further observation made is that exfoliated GO in aqueous solutions appears to be much less prone to rupture during ultrasonic treatment. Starting from identical initial graphite material, multiple micrometer large oxidized graphene flakes can still be found in the prepared solutions after sonication for significantly longer than 15 min. In contrast, chemically untreated graphene, which is subjected to the same ultrasonic bath and conditioning, quickly breaks up into much smaller elements and after 15 min it is difficult to find flakes larger than 1 μm [123]. The introduced figures show insofar representative flake sizes observed during the experiments. Chemical oxidization appears to render the carbon backbone of the graphene structure more robust.

Conclusions
In conclusion, a scalable method for the integration of single-layer chemically modified graphene and few-layer pristine graphene is investigated, with respect to three distinct solution preparation methods and dispersion principles. The most significant finding is that the solution quality is of utmost importance for the successful integration of single-layer graphene flakes. This is achieved for GO dispersed in aqueous solutions by a modified Hummers method, in contrast to chemically unmodified graphene solutions. NMP solutions yield thinner flakes deposited by DEP than SDBS stabilized solutions. Further progress must, however, be made in the liquid phase exfoliation of natural graphite to reach comparable results to GO. While two dimensional GO may be used for a large variety of sensor applications after subsequent functionalization, observed limitations in the thermal reduction process are likely to impede its use as ultrathin transparent conductors. High quality graphene solutions, largely consisting of single-layer graphene, are therefore highly anticipated for parallel graphene integration.

8 Conclusions

8.1 Results Overview

Concluding, the present thesis has made substantial contributions toward advancing the field of directed bottom-up assembly of high symmetry low-dimensional carbon nanostructures, such as carbon nanotubes and graphene, by elucidating the background of dielectrophoretic deposition and by introducing novel processes and methods for parallel device integration, as shown for piezoresistivity based pressure sensors. *General*

The framework for dielectrophoretic deposition processes is of crucial importance to understand the effects of applied inhomogeneous electric fields between electrodes when moving particles to desired locations. Three regions are distinguished by the electrical impedance measurements. In the first region the electric double layer of the electrolyte solution dominates, in the second region the conductivity of the electrolyte takes over, and in the third region the capacitive coupling of the electrodes to the chip substrate becomes the major contributing factor. The detailed analysis provides important avenues for more efficient particle handling in order to increase the dielectrophoretic deposition yield in nanostructure based networks. *DEP Framework*

The aqueous dispersion of high-quality individual SWNTs in ultra-pure long-term stable surfactant-stabilized solutions is of prime significance for directed carbon nanotube device assembly techniques. Electrical characterization of surface-synthesized SWNTs dispersed by a short ultrasound pulse and deposited by dielectrophoresis onto prefabricated electrodes evidences the high quality SWNT raw material, solution, and contact interface, important for subsequent device and sensor integration. *SWNT DEP*

Using these solutions, the selective dielectrophoretic integration of metallic SWNTs is achieved and confirmed by direct electric transport measurements at deposition frequencies above 188 MHz. A surface-conductivity model is applied and a conductivity weighting factor introduced to elucidate the separation frequency dependence. Low frequency experiments and numerical simulations show that long-range nanotube transport is governed by hydrodynamic effects, while local trapping is dominated by dielectrophoretic forces. *SWNT Separation*

Ultimately, using all the above findings, the parallel SWNT based pressure senor assembly is demonstrated. The sensors show high sensitivity at a small size while *Pressure Sensor*

8 Conclusions

consuming extremely little power. Highest resolution is achieved for large diameter, low resistance devices at the off-state of small band gap SWNTs, as predicted by the model of thermally activated transport in carbon nanotubes. The directed assembly of SGS-SWNTs, enabled by their selective dielectrophoretic integration, and subsequent device processing yields robust and long term stable sensors. The scale-up of the introduced fabrication process is straight forward and is expected to entail substantial cost benefits, which may provide a viable avenue for SWNT sensor commercialization.

Graphene DEP
In the end, the dielectrophoretic integration of single- and few-layered graphenes is introduced. To achieve single-layer pristine graphene dielectrophoretic deposition, higher solution qualities must be available, consisting largely of single-layer graphene sheets. But the reported research is expected facilitate progress toward the large-scale application of graphene-based devices.

Miniaturization
All the above accomplishments will benefit the continuous commercially driven miniaturization efforts. The directed assembly of novel low-dimensional materials has been shown, as well as refined, allowing new approaches to counter the market forces of cost reduction, device functionality and energy efficiency. Low cost, solution based bottom-up technologies are used to assemble functional transducer elements up to 3 orders of magnitude smaller than the current state of the art while all along enhancing sensor resolution and dramatically reducing drive current requirements.

8.2 Outlook

In the future, related studies may focus on various different aspects.

Solution Quality
One way to improve the dielectrophoretic integration yield is to further enhance the quality of the dispersed solutions. By employing different solvents which better match the respective surface energies of graphene (or carbon nanotubes) and the liquid, higher concentration (single-layer) solutions can be achieved and their long-term stability guaranteed [148]. The surfactant-free dispersion of carbon nanotubes may even be possible, albeit not in aqueous solutions. The solvents also have an influence on the Clausius-Mossotti factor in the dielectrophoretic force formulation, which may increase in magnitude with a prudent choice of solvent.

Monodisperse Solutions
Alternatively, recent advances in the preparation of monodisperse single-walled carbon nanotube and graphene dispersions may be used to dielectrophoretically integrate monodisperse samples. Density gradient centrifugation in conjunction with a mixture of surfactants is one possible technique to prepare these solutions [56, 149]. The dielectrophoretic assembly of small-diameter semiconducting SWNTs using these dispersions has recently been shown, something not directly possible by

8.2 Outlook

dielectophoretic deposition from heterogeneous solutions [150].

By advancing the dielectrophoretic deposition technique, the viability of the method is further strengthened for large-scale commercial interest. Latest studies have demonstrated the high mobility and on-state conductance of solution processed carbon nanotube transistors assembled by dielectrophoresis [151], low defect density in the same samples by low-temperature transport spectroscopy [115], as well as device improvements in the dielectrophoretic fabrication of chemically reduced graphene oxide field effect transistors [152]. — DEP Progress

The emergence of graphene has opened up a whole new field of application opportunities which must be carefully evaluated. Depending on the different applications of biosensors, for example, either carbon nanotubes or graphene may be the material of choice, due to their different properties [153]. No material can be said to be across-the-board better than the other. Also, early application hopes and expectations must be treated with care, as exemplified by the potential use of graphene for transparent conductors. Without doping, fundamental limitations appear to inhibit low sheet resistance in non-suspended samples [154]. — Graphene Applications

A more far-reaching outlook may see the application of the introduced techniques in the integration of semiconductor nanowires [155] and stretchable electronics [156], both research fields of great current interest. The explicit focus on applications, sensors and devices must, however, never be lost. — Future Outlook

Bibliography

[1] R. P. Feynman, "There's plenty of room at the bottom," *Engineering and Science*, pp. 22–36, February 1960.

[2] G. Binning, H. Rohrer, C. Gerber, and E. Weibel, "Surface studies by scanning tunneling microscopy," *Physical Review Letters*, vol. 49, pp. 57–61, 1982.

[3] D. M. Eigler and E. K. Schweizer, "Positioning single atoms with a scanning tunneling microscope," *Nature*, vol. 344, pp. 524–526, 1990.

[4] G. E. Moore, "Cramming more components onto integrated circuits," *Electronics*, vol. 38, 1965.

[5] J. Bardeen and W. H. Brattain, "The transistor, a semi-conductor triode," *Physical Review*, vol. 74, pp. 230–231, 1948.

[6] W. Shockley, "The theory of p-n junctions in semiconductors and p-n junction transistors," *Bell System Technical Journal*, vol. 28, pp. 435–489, 1949.

[7] D. M. Chapin, C. S. Fuller, and G. L. Pearson, "A new silicon p-n junction photocell for converting solar radiation into electrical power," *Journal of Applied Physics*, vol. 25, pp. 676–677, 1954.

[8] D. Kahng and M. M. Atalla, "Silicon-silicon dioxide surface device," in *IRE Device Research Conference*, 1960.

[9] J. S. Kilby, "Invention of the integrated circuit," U. S. Patent 3 138 743, 1964.

[10] R. N. Noyce, "Semiconductor device-and-lead structure," U. S. Patent 2 981 877, 1961.

[11] H. A. Pohl, "The motion and precipitation of suspensoids in divergent electric fields," *Journal of Applied Physics*, vol. 22, pp. 869–871, 1951.

[12] ——, *Dielectrophoresis*. Cambridge University Press, Cambridge, 1978.

[13] T. B. Jones, *Electromechanics of Particles*. Cambridge University Press, Cambridge, 1995.

Bibliography

[14] H. Morgan and N. G. Green, *AC Electrokinetics: Colloids and Nanoparticles*. Research Studies Press, Baldock, 2003.

[15] H. W. Kroto, J. R. Heath, S. C. O'Brien, R. F. Curl, and R. E. Smalley, "C_{60}: Buckminsterfullerene," *Nature*, vol. 318, pp. 162–163, 1985.

[16] S. Iijima and T. Ichihashi, "Single-shell carbon nanotubes of 1-nm diameter," *Nature*, vol. 363, pp. 603–605, 1993.

[17] D. S. Bethune, C. H. Kiang, M. S. Devries, G. Gorman, R. Savoy, J. Vazquez, and R. Beyers, "Cobalt-catalyzed growth of carbon nanotubes with single-atomic-layerwalls," *Nature*, vol. 363, pp. 605–607, 1993.

[18] R. Saito, G. Dresselhaus, and M. S. Dresselhaus, *Physical Properties of Carbon Nanotubes*. Imperial College Press, London, 1998.

[19] A. Jorio, G. Dresselhaus, and M. S. Dresselhaus, *Carbon Nanotubes: Advanced Topics in the Synthesis, Structure, Properties and Applications*. Springer, Berlin, 2008.

[20] S. Reich, C. Thomsen, and J. Maultzsch, *Carbon Nanotubes - Basic Concepts and Physical Properties*. Wiley-VCH, Weinheim, 2004.

[21] M. J. O'Connell, Ed., *Carbon Nanotubes - Properties and Applications*. Taylor & Francis, Boca Ranton, 2006.

[22] C. Dekker, "Carbon nanotubes as molecular quantum wires," *Physics Today*, vol. 52, pp. 22–28, 1999.

[23] P. L. McEuen, "Single-wall carbon nanotubes," *Physics World*, vol. 13, pp. 31–36, 2000.

[24] K. S. Novoselov, A. K. Geim, S. V. Morozov, D. Jiang, Y. Zhang, S. V. Dubonos, I. V. Grigorieva, and A. A. Firsov, "Electric field effect in atomically thin carbon films," *Science*, vol. 306, pp. 666–669, 2004.

[25] A. K. Geim and K. S. Novoselov, "The rise of graphene," *Nature Materials*, vol. 6, pp. 183–191, 2007.

[26] A. H. Castro Neto, F. Guinea, N. M. R. Peres, K. S. Novoselov, and A. K. Geim, "The electronic properties of graphene," *Reviews of modern physics*, vol. 81, pp. 109–162, 2009.

[27] A. K. Geim, "Graphene: status and prospects," *Science*, vol. 324, pp. 1530–1534, 2009.

[28] H. G. Craighead, "Nanoelectromechanical systems," *Science*, vol. 290, pp. 1532–1535, 2000.

[29] P. Avouris, Z. Chen, and V. Perebeinos, "Carbon-based electronics," *Nature Nanotechnology*, vol. 2, pp. 605–615, 2007.

[30] T. Schwamb, T.-Y. Choi, N. Schirmer, N. R. Bieri, B. Burg, J. Tharian, U. Sennhauser, and D. Poulikakos, "A dielectrophoretic method for high yield deposition of suspended, individual carbon nanotubes with four-point electrode contact," *Nano Letters*, vol. 7, pp. 3633–3638, 2007.

[31] P. A. Smith, C. D. Nordquist, T. N. Jackson, T. S. Mayer, B. R. Martin, J. Mbindyo, and T. E. Mallouk, "Electric-field assisted assembly and alignment of metallic nanowires," *Applied Physics Letters*, vol. 77, pp. 1399–1401, 2000.

[32] R. Krupke, F. Hennrich, H. B. Weber, M. M. Kappes, and H. v. Loehneysen, "Simultaneous deposition of metallic bundles of single-walled carbon nanotubes using ac-dielectrophoresis," *Nano Letters*, vol. 3, pp. 1019–1023, 2003.

[33] J. Kang, S. Hong, Y. Kim, and S. Baik, "Controlling the carbon nanotube-to-medium conductivity ratio for dielectrophoretic separation," *Langmuir*, vol. 25, pp. 12 471–12 474, 2009.

[34] B. R. Burg, J. Schneider, M. Muoth, L. Durrer, T. Helbling, N. C. Schirmer, T. Schwamb, C. Hierold, and D. Poulikakos, "Aqueous dispersion and dielectrophoretic assembly of individual surface-synthesized single-walled carbon nanotubes," *Langmuir*, vol. 25, pp. 7778–7782, 2009.

[35] R. Krupke, F. Hennrich, H. v. Loehneysen, and M. M. Kappes, "Separation of metallic from semiconducting single-walled carbon nanotubes," *Science*, vol. 301, pp. 344–347, 2003.

[36] M. Dimaki and P. Boggild, "Dielectrophoresis of carbon nanotubes using microelectrodes: a numerical study," *Nanotechnology*, vol. 15, pp. 1095–1102, 2004.

[37] D. F. Chen and H. Du, "Simulation studies on electrothermal fluid flow induced in a dielectrophoretic microelectrode system," *Journal of Micromechanics and Microengineering*, vol. 16, pp. 2411–2419, 2006.

[38] Y. Lin, J. Shiomi, S. Maruyama, and G. Amberg, "Electrothermal flow in dielectrophoresis of single-walled carbon nanotubes," *Physical Review B*, vol. 76, p. 045419, 2007.

Bibliography

[39] H. A. Haus and J. R. Melcher, *Electromagnetic Fields and Energy*. Prentice-Hall Int. Ed., London, 1989.

[40] N. G. Green, A. Ramos, and H. Morgan, "Numerical solution of the dielectrophoretic and travelling wave forces for interdigitated electrode arrays using the finite element method," *Journal of Electrostatics*, vol. 56, pp. 235–254, 2002.

[41] A. Ramos, H. Morgan, N. G. Green, and A. Castellanos, "Ac electrokinetics: a review of forces in microelectrode structures," *Journal of Physics D - Applied Physics*, vol. 31, pp. 2338–2353, 1998.

[42] N. G. Green, A. Ramos, A. Gonzalez, H. Morgan, and A. Castellanos, "Fluid flow induced by nonuniform ac electric fields in electrolytes on microelectrodes. III. Observation of streamlines and numerical simulation," *Physical Review E*, vol. 66, p. 026305, 2002.

[43] R. F. Probestein, *Physicochemical Hydrodynamics*. Wiley Inter-Science, Hoboken, 2003.

[44] N. G. Green, A. Ramos, A. Gonzalez, A. Castellanos, and H. Morgan, "Electrothermally induced fluid flow on microelectrodes," *Journal of Electrostatics*, vol. 53, pp. 71–87, 2001.

[45] D. R. Lide, *CRC Handbook of Chemistry and Physics*, 90th ed, Ed. CRC Press, London, 2009.

[46] A. Ramos, H. Morgan, N. G. Green, and A. Castellanos, "AC electric-field-induced fluid flow in microelectrodes," *Journal of Colloid and Interface Science*, vol. 217, pp. 420–422, 1999.

[47] N. G. Green, A. Ramos, A. Gonzalez, H. Morgan, and A. Castellanos, "Fluid flow induced by nonuniform ac electric fields in electrolytes on microelectrodes. I. Experimental measurements," *Physical Review E*, vol. 61, pp. 4011–4018, 2000.

[48] B. R. Burg, J. Schneider, V. Bianco, N. C. Schirmer, and D. Poulikakos, "Selective parallel integration of individual metallic single-walled carbon nanotubes from heterogeneous solutions," *Langmuir*, vol. 26, pp. 10 419–10 424, 2010.

[49] P. Somasundaran, D. W. Fuerstenau, and T. W. Healy, "Surfactant adsorption at solid-liquid interface - Dependence of mechanism on chain length," *Journal of Physical Chemistry*, vol. 68, pp. 3562–3566, 1964.

Bibliography

[50] A. Castellanos, A. Ramos, A. Gonzalez, N. G. Green, and H. Morgan, "Electrohydrodynamics and dielectrophoresis in microsystems: scaling laws," *Journal of Physics D - Applied Physics*, vol. 36, pp. 2584–2597, 2003.

[51] B. R. Burg, F. Lütolf, J. Schneider, N. C. Schirmer, T. Schwamb, and D. Poulikakos, "High-yield dielectrophoretic assembly of two-dimensional graphene nanostructures," *Applied Physics Letters*, vol. 94, p. 053110, 2009.

[52] B. R. Burg, J. Schneider, S. Maurer, N. C. Schirmer, and D. Poulikakos, "Dielectrophoretic integration of single- and few-layer graphenes," *Journal of Applied Physics*, vol. 107, p. 034302, 2010.

[53] M. J. O'Connell, S. M. Bachilo, C. B. Huffman, V. C. Moore, M. S. Strano, E. H. Haroz, K. L. Rialon, P. J. Boul, W. H. Noon, C. Kittrell, J. P. Ma, R. H. Hauge, R. B. Weisman, and R. E. Smalley, "Band gap fluorescence from individual single-walled carbon nanotubes," *Science*, vol. 297, pp. 593–596, 2002.

[54] M. F. Islam, E. Rojas, D. M. Bergey, A. T. Johnson, and A. G. Yodh, "High weight fraction surfactant solubilization of single-wall carbon nanotubes in water," *Nano Letters*, vol. 3, pp. 269–273, 2003.

[55] V. C. Moore, M. S. Strano, E. H. Haroz, R. H. Hauge, R. E. Smalley, J. Schmidt, and Y. Talmon, "Individually suspended single-walled carbon nanotubes in various surfactants," *Nano Letters*, vol. 3, pp. 1379–1382, 2003.

[56] M. C. Hersam, "Progress towards monodisperse single-walled carbon nanotubes," *Nature Nanotechnology*, vol. 3, pp. 387–394, 2008.

[57] S. Iijima, "Helical microtubules of graphitic carbon," *Nature*, vol. 354, pp. 56–58, 1991.

[58] S. M. Bachilo, L. Balzano, J. E. Herrera, F. Pompeo, D. E. Resasco, and R. B. Weisman, "Narrow (n,m)-distribution of single-walled carbon nanotubes grown using a solid supported catalyst," *Journal of the American Chemical Society*, vol. 125, pp. 11 186–11 187, 2003.

[59] S. Banerjee, T. Hemraj-Benny, and S. S. Wong, "Covalent surface chemistry of single-walled carbon nanotubes," *Advanced Materials*, vol. 17, pp. 17–29, 2005.

[60] K. Yamamoto, S. Akita, and Y. Nakayama, "Orientation of carbon nanotubes using electrophoresis," *Japanese Journal of Applied Physics, Part 2*, vol. 35, pp. L917–L918, 1996.

Bibliography

[61] T. Y. Choi, D. Poulikakos, J. Tharian, and U. Sennhauser, "Measurement of thermal conductivity of individual multiwalled carbon nanotubes by the 3-omega method," *Applied Physics Letters*, vol. 87, p. 013108, 2005.

[62] T.-Y. Choi, D. Poulikakos, J. Tharian, and U. Sennhauser, "Measurement of the thermal conductivity of individual carbon nanotubes by the four-point three-omega method," *Nano Letters*, vol. 6, pp. 1589–1593, 2006.

[63] L. Durrer, T. Helbling, C. Zenger, A. Jungen, C. Stampfer, and C. Hierold, "SWNT growth by CVD on Ferritin-based iron catalyst nanoparticles towards CNT sensors," *Sensors and Actuators B - Chemical*, vol. 132, pp. 485–490, 2008.

[64] F. Banhart, "Irradiation effects in carbon nanostructures," *Reports on Progress in Physics*, vol. 62, pp. 1181–1221, 1999.

[65] F. Hennrich, R. Krupke, K. Arnold, J. A. Rojas Stuetz, S. Lebedkin, T. Koch, T. Schimmel, and M. M. Kappes, "The mechanism of cavitation-induced scission of single-walled carbon nanotubes," *Journal of Physical Chemistry B*, vol. 111, pp. 1932–1937, 2007.

[66] A. Jungen, C. Stampfer, L. Durrer, T. Helbling, and C. Hierold, "Amorphous carbon contamination monitoring and process optimization for single-walled carbon nanotube integration," *Nanotechnology*, vol. 18, p. 075603, 2007.

[67] A. Maiti, A. Svizhenko, and M. P. Anantram, "Electronic transport through carbon nanotubes: effects of structural deformation and tube chirality," *Physical Review Letters*, vol. 88, p. 126805, 2002.

[68] A. Vijayaraghavan, S. Blatt, D. Weissenberger, M. Oron-Carl, F. Hennrich, D. Gerthsen, H. Hahn, and R. Krupke, "Ultra-large-scale directed assembly of single-walled carbon nanotube devices," *Nano Letters*, vol. 7, pp. 1556–1560, 2007.

[69] W. Kim, A. Javey, R. Tu, J. Cao, Q. Wang, and H. J. Dai, "Electrical contacts to carbon nanotubes down to 1 nm in diameter," *Applied Physics Letters*, vol. 87, p. 173101, 2005.

[70] A. Javey, J. Guo, Q. Wang, M. Lundstrom, and H. J. Dai, "Ballistic carbon nanotube field-effect transistors," *Nature*, vol. 424, pp. 654–657, 2003.

[71] R. H. Baughman, A. A. Zakhidov, and W. A. de Heer, "Carbon nanotubes - the route toward applications," *Science*, vol. 297, pp. 787–792, 2002.

[72] S. K. Hait, P. R. Majhi, A. Blume, and S. P. Moulik, "A critical assessment of micellization of sodium dodecyl benzene sulfonate (SDBS) and its interaction with poly(vinyl pyrrolidone) and hydrophobically modified polymers, JR 400 and LM 200," *Journal of Physical Chemistry B*, vol. 107, pp. 3650–3658, 2003.

[73] L. Durrer, J. Greenwald, T. Helbling, M. Muoth, R. Riek, and C. Hierold, "Narrowing SWNT diameter distribution using size-separated ferritin-based Fe catalysts," *Nanotechnology*, vol. 20, p. 355601, 2009.

[74] S. J. Tans, A. R. M. Verschueren, and C. Dekker, "Room-temperature transistor based on a single carbon nanotube," *Nature*, vol. 393, pp. 49–52, 1998.

[75] R. Martel, T. Schmidt, H. R. Shea, T. Hertel, and P. Avouris, "Single- and multi-wall carbon nanotube field-effect transistors," *Applied Physics Letters*, vol. 73, pp. 2447–2449, 1998.

[76] Z. B. Zhang, J. Cardenas, E. E. B. Campbell, and S. L. Zhang, "Reversible surface functionalization of carbon nanotubes for fabrication of field-effect transistors," *Applied Physics Letters*, vol. 87, p. 043110, 2005.

[77] T. Helbling, C. Hierold, C. Roman, L. Durrer, M. Mattmann, and V. M. Bright, "Long term investigations of carbon nanotube transistors encapsulated by atomic-layer-deposited Al2O3 for sensor applications," *Nanotechnology*, vol. 20, p. 434010, 2009.

[78] R. Saito, M. Fujita, G. Dresselhaus, and M. S. Dresselhaus, "Electronic-structure of chiral graphene tubules," *Applied Physics Letters*, vol. 60, pp. 2204–2206, 1992.

[79] H. Morgan and N. G. Green, "Dielectrophoretic manipulation of rod-shaped viral particles," *Journal of Electrostatics*, vol. 42, pp. 279–293, 1997.

[80] L. X. Benedict, S. G. Louie, and M. L. Cohen, "Static polarizabilities of single-wall carbon nanotubes," *Physical Review B*, vol. 52, pp. 8541–8549, 1995.

[81] R. Krupke, F. Hennrich, M. M. Kappes, and H. v. Loehneysen, "Surface conductance induced dielectrophoresis of semiconducting single-walled carbon nanotubes," *Nano Letters*, vol. 4, pp. 1395–1399, 2004.

[82] S. Hong, S. Jung, J. Choi, Y. Kim, and S. Baik, "Electrical transport characteristics of surface-conductance-controlled, dielectrophoretically separated single-walled carbon nanotubes," *Langmuir*, vol. 23, pp. 4749–4752, 2007.

Bibliography

[83] C. T. O'Konski, "Electric properties of macromolecules. V. Theory of ionic polarization in polyelectrolytes," *The Journal of Physical Chemistry*, vol. 64, pp. 605–619, 1960.

[84] J. J. Lyklema, *Fundamentals of Interface and Colloid Science*. Academic press, London, 1995, vol. 2.

[85] J. Lyklema and M. Minor, "On surface conduction and its role in electrokinetics," *Colloids and Surfaces A - Physicochemical and Engineering Aspects*, vol. 140, pp. 33–41, 1998.

[86] I. Ermolina and H. Morgan, "The electrokinetic properties of latex particles: comparison of electrophoresis and dielectrophoresis," *Journal of Colloid and Interface Science*, vol. 285, pp. 419–428, 2005.

[87] B. White, S. Banerjee, S. O'Brien, N. J. Turro, and I. P. Herman, "Zeta-potential measurements of surfactant-wrapped individual single-walled carbon nanotubes," *Journal of Physical Chemistry C*, vol. 111, pp. 13 684–13 690, 2007.

[88] Y. Kim, S. Hong, S. Jung, M. S. Strano, J. Choi, and S. Baik, "Dielectrophoresis of surface conductance modulated single-walled carbon nanotubes using catanionic surfactants," *Journal of Physical Chemistry B*, vol. 110, pp. 1541–1545, 2006.

[89] H. Kang, A. Kulkarni, S. Stankovich, R. S. Ruoff, and S. Baik, "Restoring electrical conductivity of dielectrophoretically assembled graphite oxide sheets by thermal and chemical reduction techniques," *Carbon*, vol. 47, pp. 1520–1525, 2009.

[90] R. Krupke, S. Linden, M. Rapp, and F. Hennrich, "Thin films of metallic carbon nanotubes prepared by dielectrophoresis," *Advanced Materials*, vol. 18, pp. 1468–1470, 2006.

[91] D. H. Shin, J.-E. Kim, H. C. Shim, J.-W. Song, J.-H. Yoon, J. Kim, S. Jeong, J. Kang, S. Baik, and C.-S. Han, "Continuous extraction of highly pure metallic single-walled carbon nanotubes in a microfluidic channel," *Nano Letters*, vol. 8, pp. 4380–4385, 2008.

[92] B. R. Burg, V. Bianco, J. Schneider, and D. Poulikakos, "Electrokinetic framework of dielectrophoretic deposition devices," *Journal of Applied Physics*, vol. 107, p. 124308, 2010.

Bibliography

[93] Z. B. Zhang, S. L. Zhang, and E. E. B. Campbell, "Dielectrophoretic behavior of ionic surfactant-solubilized carbon nanotubes," *Chemical Physics Letters*, vol. 421, pp. 11–15, 2006.

[94] A. A. Green, M. C. Duch, and M. C. Hersam, "Isolation of single-walled carbon nanotube enantiomers by density differentiation," *Nano Research*, vol. 2, pp. 69–77, 2009.

[95] J. Fraden, *Handbook of Modern Sensors: Physics, Designs, and Applications*. Springer, New York, 2004.

[96] K. E. Petersen, "Silicon as a mechanical material," *Proceedings of the IEEE*, vol. 70, pp. 420–457, 1982.

[97] W. P. Eaton and J. H. Smith, "Micromachined pressure sensors: Review and recent developments," *Smart Materials & Structures*, vol. 6, pp. 530–539, 1997.

[98] C. Roman, T. Helbling, and C. Hierold, *Springer Handbook of Nanotechnology*, B. Bhushan, Ed. Springer, Berlin, 2010.

[99] E. D. Minot, Y. Yaish, V. Sazonova, J. Y. Park, M. Brink, and P. L. McEuen, "Tuning carbon nanotube band gaps with strain," *Physical Review Letters*, vol. 90, p. 156401 , 2003.

[100] T. W. Tombler, C. W. Zhou, L. Alexseyev, J. Kong, H. J. Dai, L. Lei, C. S. Jayanthi, M. J. Tang, and S. Y. Wu, "Reversible electromechanical characteristics of carbon nanotubes under local-probe manipulation," *Nature*, vol. 405, pp. 769–772, 2000.

[101] J. Cao, Q. Wang, and H. J. Dai, "Electromechanical properties of metallic, quasimetallic, and semiconducting carbon nanotubes under stretching," *Physical Review Letters*, vol. 90, p. 157601, 2003.

[102] R. J. Grow, Q. Wang, J. Cao, D. W. Wang, and H. J. Dai, "Piezoresistance of carbon nanotubes on deformable thin-film membranes," *Applied Physics Letters*, vol. 86, p. 093104, 2005.

[103] C. Stampfer, T. Helbling, D. Obergfell, B. Schoberle, M. K. Tripp, A. Jungen, S. Roth, V. M. Bright, and C. Hierold, "Fabrication of single-walled carbon-nanotube-based pressure sensors," *Nano Letters*, vol. 6, pp. 233–237, 2006.

[104] C. Stampfer, A. Jungen, R. Linderman, D. Obergfell, S. Roth, and C. Hierold, "Nano-electromechanical displacement sensing based on single-walled carbon nanotubes," *Nano Letters*, vol. 6, pp. 1449–1453, 2006.

Bibliography

[105] T. Helbling, C. Roman, L. Durrer, C. Stampfer, and C. Hierold, "Gauge factor tuning and long term stability of nano electromechanical carbon nanotube sensors," *Submitted for publication*, 2010.

[106] T. Helbling, C. Roman, and C. Hierold, "Signal-to-noise ratio in carbon nanotube electromechanical piezoresistive sensors," *Nano Letters*, vol. 10, 2010.

[107] A. Kleiner and S. Eggert, "Band gaps of primary metallic carbon nanotubes," *Physical Review B*, vol. 63, p. 073408, 2001.

[108] S. P. Timoshenko and S. Woinowsky-Krieger, *Theory of Plates and Shells*. McGraw-Hill, Auckland, 1954.

[109] E. Ventsel and T. Krauthammer, *Thin Plates and Shells - Theory, Analysis and Applications*. Marcel Dekker, New York, 2001.

[110] H. J. Timme, *Advanced Micro and Nanosystems: CMOS - MEMS*, H. Baltes, O. Brand, K. Fedder, C. Hierold, J. Korvink, and O. Tabata, Eds. Wiley-VCH, Weinheim, 2005, vol. 2.

[111] C. W. Zhou, J. Kong, and H. J. Dai, "Electrical measurements of individual semiconducting single-walled carbon nanotubes of various diameters," *Applied Physics Letters*, vol. 76, pp. 1597–1599, 2000.

[112] Z. H. Chen, J. Appenzeller, J. Knoch, Y. M. Lin, and P. Avouris, "The role of metal-nanotube contact in the performance of carbon nanotube field-effect transistors," *Nano Letters*, vol. 5, pp. 1497–1502, 2005.

[113] D. Mann, A. Javey, J. Kong, Q. Wang, and H. J. Dai, "Ballistic transport in metallic nanotubes with reliable Pd ohmic contacts," *Nano Letters*, vol. 3, pp. 1541–1544, 2003.

[114] T. Helbling, "Carbon nanotube field effect transistors as electromechanical transducers," Ph.D. dissertation, ETH Zurich, 2010.

[115] P. Stokes and S. I. Khondaker, "Evaluating defects in solution-processed carbon nanotube devices via low-temperature transport spectroscopy," *ACS Nano*, vol. 4, pp. 2659–2666, 2010.

[116] S. Stankovich, D. A. Dikin, G. H. B. Dommett, K. M. Kohlhaas, E. J. Zimney, E. A. Stach, R. D. Piner, S. T. Nguyen, and R. S. Ruoff, "Graphene-based composite materials," *Nature*, vol. 442, pp. 282–286, 2006.

Bibliography

[117] X. Wang, L. Zhi, and K. Muellen, "Transparent, conductive graphene electrodes for dye-sensitized solar cells," *Nano Letters*, vol. 8, pp. 323–327, 2008.

[118] J. Wu, H. A. Becerril, Z. Bao, Z. Liu, Y. Chen, and P. Peumans, "Organic solar cells with solution-processed graphene transparent electrodes," *Applied Physics Letters*, vol. 92, p. 263302, 2008.

[119] S. Stankovich, D. A. Dikin, R. D. Piner, K. A. Kohlhaas, A. Kleinhammes, Y. Jia, Y. Wu, S. T. Nguyen, and R. S. Ruoff, "Synthesis of graphene-based nanosheets via chemical reduction of exfoliated graphite oxide," *Carbon*, vol. 45, pp. 1558–1565, 2007.

[120] X. Li, X. Wang, L. Zhang, S. Lee, and H. Dai, "Chemically derived, ultrasmooth graphene nanoribbon semiconductors," *Science*, vol. 319, pp. 1229–1232, 2008.

[121] Y. Si and E. T. Samulski, "Synthesis of water soluble graphene," *Nano Letters*, vol. 8, pp. 1679–1682, 2008.

[122] X. Li, G. Zhang, X. Bai, X. Sun, X. Wang, E. Wang, and H. Dai, "Highly conducting graphene sheets and Langmuir-Blodgett films," *Nature Nanotechnology*, vol. 3, pp. 538–542, 2008.

[123] Y. Hernandez, V. Nicolosi, M. Lotya, F. M. Blighe, Z. Sun, S. De, I. T. McGovern, B. Holland, M. Byrne, Y. K. Gun'ko, J. J. Boland, P. Niraj, G. Duesberg, S. Krishnamurthy, R. Goodhue, J. Hutchison, V. Scardaci, A. C. Ferrari, and J. N. Coleman, "High-yield production of graphene by liquid-phase exfoliation of graphite," *Nature Nanotechnology*, vol. 3, pp. 563–568, 2008.

[124] H. A. Pohl and I. Hawk, "Separation of living and dead cells by dielectrophoresis," *Science*, vol. 152, pp. 647–649, 1966.

[125] M. Washizu and O. Kurosawa, "Electrostatic manipulation of DNA in microfabricated structures," *IEEE Transactions on industry applications*, vol. 26, pp. 1165–1172, 1990.

[126] X. Q. Chen, T. Saito, H. Yamada, and K. Matsushige, "Aligning single-wall carbon nanotubes with an alternating-current electric field," *Applied Physics Letters*, vol. 78, pp. 3714–3716, 2001.

[127] S. Hong, S. Jung, S. Kang, Y. Kim, X. Chen, S. Stankovich, S. R. Ruoff, and S. Baik, "Dielectrophoretic deposition of graphite oxide soot particles," *Journal of Nanoscience and Nanotechnology*, vol. 8, pp. 424–427, 2008.

Bibliography

[128] W. S. Hummers and R. E. Offeman, "Preparation of graphitic oxide," *Journal of the American Chemical Society*, vol. 80, p. 1339, 1958.

[129] M. Hirata, T. Gotou, S. Horiuchi, M. Fujiwara, and M. Ohba, "Thin-film particles of graphite oxide 1: High-yield synthesis and flexibility of the particles," *Carbon*, vol. 42, pp. 2929–2937, 2004.

[130] G. Titelman, V. Gelman, S. Bron, R. Khalfin, Y. Cohen, and H. Bianco-Peled, "Characteristics and microstructure of aqueous colloidal dispersions of graphite oxide," *Carbon*, vol. 43, pp. 641–649, 2005.

[131] H. A. Becerril, J. Mao, Z. Liu, R. M. Stoltenberg, Z. Bao, and Y. Chen, "Evaluation of solution-processed reduced graphene oxide films as transparent conductors," *ACS Nano*, vol. 2, pp. 463–470, 2008.

[132] G. Eda, G. Fanchini, and M. Chhowalla, "Large-area ultrathin films of reduced graphene oxide as a transparent and flexible electronic material," *Nature Nanotechnology*, vol. 3, pp. 270–274, 2008.

[133] N. I. Kovtyukhova, P. J. Ollivier, B. R. Martin, T. E. Mallouk, S. A. Chizhik, E. V. Buzaneva, and A. D. Gorchinskiy, "Layer-by-layer assembly of ultrathin composite films from micron-sized graphite oxide sheets and polycations," *Chemistry of Materials*, vol. 11, pp. 771–778, 1999.

[134] S. Park and R. S. Ruoff, "Chemical methods for the production of graphenes," *Nature Nanotechnology*, vol. 4, pp. 217–224, 2009.

[135] X. Wu, M. Sprinkle, X. Li, F. Ming, C. Berger, and W. A. de Heer, "Epitaxial-graphene/graphene-oxide junction: An essential step towards epitaxial graphene electronics," *Physical Review Letters*, vol. 101, p. 026801, 2008.

[136] A. Vijayaraghavan, C. Sciascia, S. Dehm, A. Lombardo, A. Bonetti, A. C. Ferrari, and R. Krupke, "Dielectrophoretic assembly of high-density arrays of individual graphene devices for rapid screening," *ACS Nano*, vol. 3, pp. 1729–1734, 2009.

[137] B. Partoens and F. M. Peeters, "From graphene to graphite: Electronic structure around the K point," *Physical Review B*, vol. 74, p. 075404, 2006.

[138] W. Gao, L. B. Alemany, L. Ci, and P. M. Ajayan, "New insights into the structure and reduction of graphite oxide," *Nature Chemistry*, vol. 1, pp. 403–408, 2009.

[139] V. C. Tung, M. J. Allen, Y. Yang, and R. B. Kaner, "High-throughput solution processing of large-scale graphene," *Nature Nanotechnology*, vol. 4, pp. 25–29, 2009.

[140] I. Jung, D. A. Dikin, R. D. Piner, and R. S. Ruoff, "Tunable electrical conductivity of individual graphene oxide sheets reduced at "low" temperatures," *Nano Letters*, vol. 8, pp. 4283–4287, 2008.

[141] M. Lotya, Y. Hernandez, P. J. King, R. J. Smith, V. Nicolosi, L. S. Karlsson, F. M. Blighe, S. De, Z. Wang, I. T. McGovern, G. S. Duesberg, and J. N. Coleman, "Liquid phase production of graphene by exfoliation of graphite in surfactant/water solutions," *Journal of the American Chemical Society*, vol. 131, pp. 3611–3620, 2009.

[142] C. Valles, C. Drummond, H. Saadaoui, C. A. Furtado, M. He, O. Roubeau, L. Ortolani, M. Monthioux, and A. Penicaud, "Solutions of negatively charged graphene sheets and ribbons," *Journal of the American Chemical Society*, vol. 130, pp. 15802–15804, 2008.

[143] S. Park, J. An, I. Jung, R. D. Piner, S. J. An, X. Li, A. Velamakanni, and R. S. Ruoff, "Colloidal suspensions of highly reduced graphene oxide in a wide variety of organic solvents," *Nano Letters*, vol. 9, pp. 1593–1597, 2009.

[144] C. E. Hamilton, J. R. Lomeda, Z. Sun, J. M. Tour, and A. R. Barron, "High-yield organic dispersions of unfunctionalized graphene," *Nano Letters*, vol. 9, pp. 3460–3462, 2009.

[145] D. Graf, F. Molitor, K. Ensslin, C. Stampfer, A. Jungen, C. Hierold, and L. Wirtz, "Spatially resolved raman spectroscopy of single- and few-layer graphene," *Nano Letters*, vol. 7, pp. 238–242, 2007.

[146] C. Gomez-Navarro, R. T. Weitz, A. M. Bittner, M. Scolari, A. Mews, M. Burghard, and K. Kern, "Electronic transport properties of individual chemically reduced graphene oxide sheets," *Nano Letters*, vol. 7, pp. 3499–3503, 2007.

[147] R. S. Sundaram, C. Gomez-Navarro, E. J. H. Lee, M. Burghard, and K. Kern, "Noninvasive metal contacts in chemically derived graphene devices," *Applied Physics Letters*, vol. 95, p. 223507, 2009.

[148] J. N. Coleman, "Liquid-phase exfoliation of nanotubes and graphene," *Advanced Functional Materials*, vol. 19, pp. 3680–3695, 2009.

Bibliography

[149] A. A. Green and M. C. Hersam, "Emerging methods for producing monodisperse graphene dispersions," *Journal of Physical Chemistry Letters*, vol. 1, pp. 544–549, 2010.

[150] A. Vijayaraghavan, F. Hennrich, N. Stuerzl, M. Engel, M. Ganzhorn, M. Oron-Carl, C. W. Marquardt, S. Dehm, S. Lebedkin, M. M. Kappes, and R. Krupke, "Toward single-chirality carbon nanotube device arrays," *ACS Nano*, vol. 4, pp. 2748–2754, 2010.

[151] P. Stokes and S. I. Khondaker, "High quality solution processed carbon nanotube transistors assembled by dielectrophoresis," *Applied Physics Letters*, vol. 96, p. 083110, 2010.

[152] D. Joung, A. Chunder, L. Zhail, and S. I. Khondaker, "High yield fabrication of chemically reduced graphene oxide field effect transistors by dielectrophoresis," *Nanotechnology*, vol. 21, p. 165202, 2010.

[153] W. Yang, K. R. Ratinac, S. P. Ringer, P. Thordarson, J. J. Gooding, and F. Braet, "Carbon nanomaterials in biosensors: Should you use nanotubes or graphene?" *Angewandte Chemie - International Edition*, vol. 49, pp. 2114–2138, 2010.

[154] S. De and J. N. Coleman, "Are there fundamental limitations on the sheet resistance and transmittance of thin graphene films?" *ACS Nano*, vol. 4, pp. 2713–2720, 2010.

[155] P. Yang, R. Yan, and M. Fardy, "Semiconductor nanowire: What's next?" *Nano Letters*, vol. 10, pp. 1529–1536, 2010.

[156] J. A. Rogers, T. Someya, and Y. Huang, "Materials and mechanics for stretchable electronics," *Science*, vol. 327, pp. 1603–1607, 2010.

Die VDM Verlagsservicegesellschaft sucht für wissenschaftliche Verlage abgeschlossene und herausragende

Dissertationen, Habilitationen, Diplomarbeiten, Master Theses, Magisterarbeiten usw.

für die kostenlose Publikation als Fachbuch.

Sie verfügen über eine Arbeit, die hohen inhaltlichen und formalen Ansprüchen genügt, und haben Interesse an einer honorarvergüteten Publikation?

Dann senden Sie bitte erste Informationen über sich und Ihre Arbeit per Email an *info@vdm-vsg.de*.

Sie erhalten kurzfristig unser Feedback!

VDM Verlagsservicegesellschaft mbH
Dudweiler Landstr. 99   Telefon +49 681 3720 174
D - 66123 Saarbrücken   Fax   +49 681 3720 1749
www.vdm-vsg.de

Die VDM Verlagsservicegesellschaft mbH vertritt

Dissertationen, Habilitationen,
Diplomarbeiten, Master Theses

Printed by Books on Demand GmbH, Norderstedt / Germany